About Island Press

Island Press, a nonprofit organization, publishes, markets, and distributes the most advanced thinking on the conservation of our natural resources—books about soil, land, water, forests, wildlife, and hazardous and toxic wastes. These books are practical tools used by public officials, business and industry leaders, natural resource managers, and concerned citizens working to solve both local and global resource problems.

Founded in 1978, Island Press reorganized in 1984 to meet the increasing demand for substantive books on all resource-related issues. Island Press publishes and distributes under its own imprint and offers these services to other nonprofit organizations.

Support for Island Press is provided by Apple Computers, Inc., The Mary Reynolds Babcock Foundation, The Educational Foundation of America, The Charles Engelhard Foundation, The Ford Foundation, The George Gund Foundation, The William and Flora Hewlett Foundation, The Joyce Foundation, The J. M. Kaplan Fund, The John D. and Catherine T. MacArthur Foundation, The Andrew W. Mellon Foundation, The Joyce Mertz-Gilmore Foundation, The New-Land Foundation, Northwest Area Foundation, The Jessie Smith Noyes Foundation, The J. N. Pew, Jr., Charitable Trust, The Rockefeller Brothers Fund, The Florence and John Schumann Foundation, The Tides Foundation, and individual donors.

About The Wilderness Society

The Wilderness Society is a nonprofit organization with more than 325,000 members devoted to preserving wilderness and wildlife, protecting America's forests, parks, rivers, wildlife refuges, and shorelands, and fostering an American land ethic. The society was founded in 1935 by a group of tireless advocates—including professional foresters Aldo Leopold and Robert Marshall.

Today the society combines the experience and skills of foresters, ecologists, economists, land-planning and resource experts, media and public education specialists, and strategists to bring the vision of its founders to fruition. In concert with thirteen field offices around the country, staff in Washington, D.C., continue the effort to conserve biological diversity, provide wildlife habitat and healthy fisheries, protect high-quality watersheds, ensure outdoor recreation opportunities, and—especially through wilderness designations—leave a legacy of scenic beauty, scientific laboratories, and places of unmatched solitude for future generations.

ANCIENT
FORESTS
of the PACIFIC
NORTHWEST

ELLIOTT A. NORSE

Foreword by Peter H. Raven

THE WILDERNESS SOCIETY

ANCIENT FORESTS of the PACIFIC NORTHWEST

ISLAND PRESS

Washington, D.C. □ *Covelo, California*

Library of Congress Cataloging-in-Publication Data

Norse, Elliott A.
Ancient forests of the Pacific Northwest / Elliott A. Norse.
p.   cm.
Includes bibliographical references.
ISBN 1-55963-017-5.—ISBN 1-55963-016-7 (pbk.)
1. Old growth forests—Northwest, Pacific.   2. Forest ecology—Northwest, Pacific.   3. Biological diversity conservation—Northwest, Pacific.   4. Sustainable forestry—Northwest, Pacific.   5. Forest conservation—Northwest, Pacific.   6. Timber—Northwest, Pacific.   I. Title.

SD387.043N67   1990   89-20029
634.9'09795—dc20   CIP

The author is grateful for permission to include the following previously copyrighted material:

Figure 6-1 on page 162 is from The Fragmented Forest: Island Biogeography Theory and the Preservation of Biotic Diversity *by Larry D. Harris, published by University of Chicago Press, 1984.*

Figure 6-2 on page 166, for reptiles and amphibians, R. H. MacArthur and E. O. Wilson, The Theory of Island Biogeography *(Princeton, N.J.: Princeton University Press, 1967), Figure 2, p. 8 (copyright 1967 by Princeton University Press; Princeton Paperback, 1967); for bird species, Jared M. Diamond,* The Island Dilemma: Lessons of Modern Biogeographic Studies for the Design of Natural Reserves *(London: Applied Science Publishers Ltd., 1975), p. 131 (copyright by Allied Science Publishers Ltd.).*

Printed on recycled, acid-free paper

Manufactured in the United States of America

10   9   8   7   6   5   4   3   2

Contents

List of Contributed Essays

List of Photographs

Preface

Life is remarkably resilient. When challenged by environmental change, individuals, species, and ecosystems have ways of adjusting. But living systems are also vulnerable to change. They have limits, that, if exceeded, cause them to decline and collapse.

Humans are exceeding the limits of a growing number of living systems. This is not something we can ignore because their loss threatens our own existence: People are part of the biosphere and utterly depend on other living things for everything from the food we eat to the oxygen we breathe. The Earth's living systems are our larders and life-support systems. To survive, we must avoid overwhelming them. One living system now being overwhelmed is the ancient forests of the Pacific Northwest.

I knew it intellectually when The Wilderness Society (TWS) asked me to write this book. I also knew that the society plays a unique and crucial role in conservation because, a few years ago, when I was Public Policy Director of the Ecological Society of America, The Wilderness Society asked me and my colleagues to write *Conserving Biological Diversity in Our National Forests* (1986). I was impressed by TWS's openness to scientists' perspectives on policy issues; few organizations advocating public policies are as careful about getting their facts straight. And our collaboration succeeded: The little "green book" has helped many people involved in the forest planning process. Therefore, I had reason to be happy when asked to write this book. But there was also a more personal reason.

I have always loved living things, and my preschool fascination with marine biota led (with hardly any digression) to doctoral and postdoctoral research on the distribution ecology of crabs in tropical marine ecosystems. But on the block where first I lived was a big, old poplar, whose shimmering leaves and cool shade left me with a special appreciation for trees.

On my first journey into an ancient forest, in 1974, the clean, moist air, the lush vegetation of unearthly scale, the nurse logs on the forest floor, and the sunbeams filtering through the canopy touched not only the scientist in me, but also something more basic. Amidst forests of Olympian majesty and beauty, I felt I had come home. But on viewing the ubiquitous, raw clearcuts on public lands ringing Olympic National Park, I felt something was very, very wrong.

I am far from alone in this feeling. Knowing that federal and state agencies are promoting the destruction of some of nature's oldest, largest living creations troubles many people. Yet most of us also realize that our society consumes large amounts of lumber and paper (including paper for books about ancient forests), and to have them, trees must fall. Fortunately, trees and even forests can be renewable. But are they being renewed? Are the people entrusted with managing our forests maintaining their values for us all?

In search of answers, I went to live in the Northwest, reading about ancient forests in newspapers, scientific journals, and books, speaking with people who know their values and vulnerabilities, and exploring them in aircraft, on wheels, on foot: looking, listening, feeling, smelling, and tasting them. And although I was humbled to learn that many crucial data simply do not exist, what I did learn is deeply troubling: that the world's finest coniferous forests are neither being protected nor exploited wisely. They are being destroyed, and quickly too.

I also found reason for hope. Many forest scientists and managers told me that Pacific Northwest forestry is changing.

A growing number of forest managers know that forests are more than just board feet. They are working to bring about the birth of a "new forestry," which, recognizing that commodity production depends on the integrity of forest ecosystems, sustains them both. They acknowledge that foresters have made serious mistakes and that changing our present destructive course must inevitably hurt some people in the short term—something none of us welcomes—but that in the long term the consequences will be far worse if we don't change course.

And this change is needed not in twenty years, not in five years, but now. Sound figures are now available, and they show that ancient forests are disappearing much faster than previously thought. Ancient forests are also facing external threats that were unimagined until recently and are still largely unappreciated. Knowing this has the potential to unite long-standing enemies—loggers and conservationists alike—for we are losing something important to us all. But to save the ancient forests—both the rich biota and strong, fine-grained, knot-free heart-wood they yield—we will have to look at them in a new way.

In hope of encouraging this new perspective, I offer timber industry officials, conservationists, agency officials, scientists, legislators, students, and informed citizens an ecologist's view of the biological values of ancient forests, how those values are being eroded, and what it will take to maintain them. The insights of scientists are crucial for managing and conserving ancient forests. Unfortunately, our insights have not been readily available to the public, the civil servants who manage our public forestlands, and the political leaders who, together, will determine their fate.

The evidence and my conclusions will undoubtedly trouble some of you. Rest assured that they trouble me too. But I ask that you examine their validity with care, for a great deal is at stake.

This book focuses on one geographic region of the United States, but its basic story—with amazingly few variations—is being repeated from the North Woods of Maine to the flatwoods of Florida, from the Brazilian rainforests of Rondonia to the Australian gum forests of Queensland. The Pacific Northwest is hardly unique. Indeed, in nearby British Columbia, ancient forests are disappearing even faster. Driven by uncontrolled population growth and escalating demands, people are destroying the resource base on which we depend.

Our only chance of achieving and sustaining a decent standard of living lies in understanding living systems. This book, then, is an attempt to equip us for maintaining the source of our wealth and well-being in one very special corner of the Earth. You might think that understanding is not enough. But I believe that if enough of us understand that these remarkable ecosystems are vital to our self-interest, we cannot help but find the ingenuity and wisdom to protect their integrity. Indeed, I would wager that people who once sold or logged them will become their staunchest and ablest defenders.

Regrettably, I did not grow up among the ancient forests. There is so much to know about their ecology, economics, and politics that, lacking lifetimes of training and experience, I have depended on information, insights, and guidance generously provided by scientists, managers, and concerned citizens, old friends and new friends, to whom I am indebted. These include Susan Abrams, Jim Agee, Harriet Allen, Melody Allen, Joe Ammiratti, Keith and Carol Aubry, Malcolm Baldwin, Dinah Bear, Doug Bertran, Rob Bierregaard, Dan Bigger, Dwight Billings, Barbara Bizilia, Larry Bliss, Bobcat, Dee Boersma, Jill Bouma, Gary Braasch,

Rick Brown, Evelyn Brownlee, Carol Bruun, Bill Burley, Bruce Bury, Andy Carey, Dean Carrier, George Carroll, Anne Chalmers, Ellen Chu, Bill Cleaver, Steve Cline, Joan Consani, Steve Corn, Len Cornelius, Rod Crawford, Tim Crosby, Terry Cundy, Virginia Dale, Bill and Margo Denison, Jim DePree, Margaret Dilly, Robert Edmonds, Keith Ervin, Steve Eubanks, Gary and Linda Fields, John Firor, Chris Fischer, Kathy Fletcher, Jerry Franklin, Tom Franklin, Terry Frest, Len Gardner, Stanley Gessel, Carol Green, Ev Hansen, Mark Harmon, Dennis Harr, Miles Hemstrom, Sandra Henderson, Vi Hilbert, Bob Hofman, Skee Houghton, William Houser, Mark Huff, Franz Jantzen, Deborah Jensen, Murray Johnson, Rolf Johnson, Carol Jolly, Tom Juelson, Andy Kerr, Neil Korman, Bill Kunin, Jack Lattin, Dave McCorkle, Jim McIver, Doug MacWilliams, Dave Manuwal, Bruce Marcot, Jina Mariani, Dave Marshall, Chris Maser, Amy Mathews, Bob Matlock, Dick Miller, Andy Moldenke, Ron Neilson, Barry Noon, Julie Norman, Ketzel Norse, Gordon Orians, Gary Parsons, Dennis Paulson, Bob Pearson, Malcolm Penny, Dave and Carol Perry, Rob Peters, Bob Pyle, Ken Raedeke, C. J. Ralph, Martin Rand, Marty Raphael, Peter Raven, Chad Roberts, Mike Robinson, Seri Rudolph, Len Ruggiero, Jerry Rust, John Schoen, Tim Schowalter, Jim Sedell, Roger Sedjo, Barbara Shapiro, Steve Shaw, Jean Siddall, Winnie Sidle, Steve Sillett, Chuck Sisco, Dick Smythe, Phil Sollins, Tom Spies, John Stanley, Linda Starke, Dave Stokes, Matthew Suffness, Fred Swanson, Steve Swartz, Mary Symmes, Robert Szaro, Dick Taber, Alan Teramura, Jack Thomas, David Thorud, Jim Trappe, Thomas Tuchmann, Dave Turner, John Twiss, Fiorenzo Ugolini, Dan Varoujean, Kristiina Vogt, Dick Waring, Lee Webb, Mike Weber, Susan Weber, Ken Weiner, Steve West, Dave Wheeler, Dave Willis, Ruth Windhover, George Woodwell, Bob Worrest, Bob Ziemer, Pete Ziminski, and Dave Zirkle.

It has been my pleasure to work with and learn from an extraordinarily competent and dedicated team, the staff and leaders of The Wilderness Society, many of whom provided assistance and astute criticism while encouraging me to say what I think needs saying. They include Mike Anderson, Pat Attkisson, Syd Butler, Claudia Carter, John Castagna, Peter Coppelman, Jean Durning, Pete Emerson, Barry Flamm, George Frampton, Bob Freimark, Pat Harris, Anna Harrison, Pat Holmes, Rich Hoppe, Debaran Kelso, Kathy Kilmer, Peter Kirby, Deanne Kloepfer, Lola McClintock, Joe Mehrkens, Peter Morrison, Senator Gaylord Nelson, Jeff Olson, Pat Reed, Bill Reffalt, Dick Rice, Melanie Rowland, Patty Schifferle, Doreen Schmidt, Steve Sekscienski, Karin

Sheldon, John Shepard, Ron Tipton, Larry Tuttle, Tom Watkins, Jay Watson, Steve Whitney, David Wilcove, and Becca Wodder.

I am most grateful to the W. Alton Jones Foundation, which funded this work, for its clear vision and unstinting commitment to fostering conservation of biological diversity. At a time when so many people are saying that something needs to be done, the Jones Foundation is *doing* something, and doing it very well.

A manuscript differs from a book in the way that a rough-hewn hunk of marble differs from a finished sculpture. Barbara Dean, Will Farnam, Chuck Savitt, Nancy Seidule, Barbara Youngblood, Estelle Jelinek, and the other people of Island Press made the process of bringing forth the book from a manuscript a pleasure that I will never forget.

Finally, I must acknowledge the contributions of three people. My uncle and namesake, Elliott Albert, died on a European battlefield before I was born, but stories of his ethics and love of nature spawned my interest in conservation. Despite having little formal education herself, my mother, Harriett Sigman, fed my insatiable appetite for books about biology long before I began school, and served abundant helpings of love and wisdom with them. And my father, Larry Norse, who crafted fur coats from the pelts of forest mammals, taught me how conflicted a person can feel when he makes his living exploiting what he knows and cares for. Without their teachings, I could not have begun this book, and so, it is to them I dedicate it.

Foreword

We live together in a world where more than five billion people are straining the productive capacity of biological systems everywhere to support us now, and prospects for the future appear even more challenging. In view of the demands for ecological moderation that are constantly made by industrialized countries to their tropical counterparts, it is sobering to realize that here in the United States we are actively engaged in sacrificing the remnants of our own ancient forests for a few years of additional profit. The unique qualities of these forests, illustrated so well in the pages that follow, are being lost, as we pay seemingly little attention to their irreplaceability and beauty. Collectively we condone this loss despite the fact that the forests contribute fundamentally to the scenic values of the areas where they occur and to the enjoyment of living in and visiting the Northwest.

Although we live in a world where the harvesting of petroleum, groundwater, trees, and fishes—stocks that have been built up over millions of years—is coming to an end, we continue to find it difficult to decide when and how to make the transition to sustainability and what can be saved in the process. These relationships imply a series of unpleasant choices—ones that many of us would prefer to avoid contemplating. Nonetheless, the choices are genuine, and we are making them, whether or not we wish to focus on the process. Meanwhile, we must ask ourselves, we who live in the United States, citizens of the wealthiest nation on Earth, how we can continue to operate in a manner seemingly so contrary to our own interests, and what lessons we can learn from our own situation that might apply to the appropriate management of the world as a whole.

In the pages of this book, Elliott Norse provides a much-needed overview of the ecology and conservation of the old-growth forests of the Pacific Northwest. By documenting with scientific rigor the reasons that we must save them—simply put, they are the remnants of the finest coniferous forests in the world—Norse offers us reason for hope and a blueprint for constructive action. Emphasizing the interaction between the species of plants, animals, fungi, and microorganisms that make up these rich and diverse forests, he helps the reader to understand how they are unique, how little we really know about them, and why we should cherish them more than we do. Left alone, these forests help to provide a clean atmosphere; they build and protect soils and watersheds; and they provide a home for a rich and varied complement of biological diversity. Their loss amounts to the loss of an irreplaceable part of our national heritage, and of values that are unique, even when measured on a global scale. This loss is as permanent as is the notorious clearing of their counterparts in the tropics, and it will have serious and lasting repercussions on the quality of life in the Pacific Northwest and the ecology of the areas where they have flourished.

By treating 500- to 1,000-year-old forests as if they were a renewable resource, we are acting out a fiction, and thereby making a grave mistake. Forests are indeed renewable, but once they have been removed from a particular area, the ancient forests described in this book will never appear again, given the nature of human activities in the contemporary world and their consequences. By clearing such forests on both public and private lands, we are therefore losing them forever on a regional scale; they may be replaced with decades-old successional forest that can indeed be lumbered continuously, but that forest is in no way—biologically, scenically, or in terms of its contribution to the quality of human existence—the equivalent of what is being lost. Indeed, all of the ancient forests that remain could be saved, with no lasting impact on the regional economies, simply by accelerating the inevitable shift of the timber industry to second-growth forest on lands that were, in many cases, cleared decades ago. Federal policy, which tends to encourage the removal of the old-growth forests first, is therefore ill informed because it is not leading to the kind of sustained use of our natural resources that was envisioned by those who established the concept of national forests in the first place. Those of us who cherish these values and hope that our children and grandchildren will be able to enjoy the same advantages that are available during our lifetimes must work to change these policies for their benefit. In a "land of many uses," the extraction of timber, a single kind of activity among many actual and potential ones, cannot be allowed to predominate to the extent that it has for many decades: under prevailing conditions, the system simply cannot last.

Throughout the tropics of the world, primary evergreen forest, more

commonly known as tropical rainforest, is being cut and burned, usually for the development of agriculture, but sometimes for lumber. The
causes of its loss are complex and are related to the explosive growth of
the human populations that live in tropical countries, the serious and
extensive poverty that afflicts such a high proportion of those people,
and their collective unwillingness to deal with the problem of conserving the forests. In addition, there is an implicit assumption that any
clearing of forests is somehow good and proper, advancing human
values; the fact that many tropical forests grow on soils too poor to
cultivate on a sustainable basis using available techniques is ignored,
and the assumption that tropical countries are simply emulating their
industrialized counterparts is used as an excuse to exhibit belligerent
nationalism rather than to examine the consequences for their future.
None of these conditions ought to be characteristic of the United States,
the richest country on Earth: when we reflect that the industrialized
nations of the world, with a rapidly shrinking quarter of the global
population, control about 85 percent of the world's wealth and are home
to some 94 percent of the world's scientists and engineers, we must
indeed wonder why we cannot conserve our own natural resources by
dealing with them in a sustainable manner, and what hope there is for
the long-term sustainability of our planetary home if we cannot muster
the will to do so. If we continue to present the kind of model that our
actions clearly indicate, how can we hope to provide world leadership in
conservation? If we lack the will to do so, who will fill the void?

Those of us who have long been involved in the struggle to sustain and
better the lot of the people who live in the tropics—both for humane
reasons and because we have concluded that only by improving their
welfare can we collectively make possible the long-term preservation
of the unique values of their environment—understand very well the
role that our limited knowledge of other nations plays in determining
our response to their plight. While working to improve that situation,
we simultaneously observe what a tragedy it is that we do not understand our own forests well enough to love and preserve them. Wood and
other products derived from trees will be required indefinitely, as far as
anyone can tell, but, like all other products, they ultimately must be
produced on a sustainable basis. For the Pacific Northwest and other
regions with ancient, primary forests, that means that the timber industry must shift from one-time "mining" of forests that will never grow
again to the permanent use of forests that are replanted and harvested
like crops. The responsible attainment of this goal over wide areas has

made the United States timber industry a much-appreciated world model, but it remains a national tragedy that we still demand the right to use up, over the next five years or so, such a large proportion of the unique old-growth forests that Norse describes so lovingly here. Unless we care enough to change our policies, the last unprotected stands of these magnificent forests, of which only about an eighth remains uncut, are likely to be irreparably fragmented before the end of the century.

The values of Arne Naess, which many of us share, inspire this book, and are expressed well in the following lines:

> ... biologists have the precious privilege to be acquainted with worlds, largely outside the experience of others—worlds of microscopic living beings, and of life processes which amaze us all. These are sources of joy and wonder that the biologist should be able to expose and communicate, not only in the form of textbooks, but also through the direct language of spontaneous experience. . . . These must not be left unexpressed in the name of science. There may of course be biologists who suspect that an expert who uses spontaneous language is incapable of scientific rigor. But we should not neglect our linguistic abilities from fear of such misunderstandings. When biologists refrain from using the rich and flavorful language of their own spontaneous experience of all life forms—not only of the spectacularly beautiful but also of the mundane and bizarre as well—they support the value nihilism which is implicit in outrageous environmental policies. (Arne Naess, "Intrinsic Value: Will the Defenders of Nature Please Rise?" in M. E. Soulé, ed., *Conservation Biology*, 1988, p. 512.)

For his successful emulation of these values, we owe a debt of gratitude to Elliott Norse, as we do to The Wilderness Society for sponsoring the work and to the W. Alton Jones Foundation, which has played such a signal role in creatively promoting the conservation of biological diversity throughout the world, for sponsoring the project. I believe that this effort constitutes a lasting contribution to the preservation of the ancient forests—for us and for future generations to enjoy—and I have no higher compliment to bestow.

Peter H. Raven
Director
Missouri Botanical Garden

ANCIENT
FORESTS
of the PACIFIC
NORTHWEST

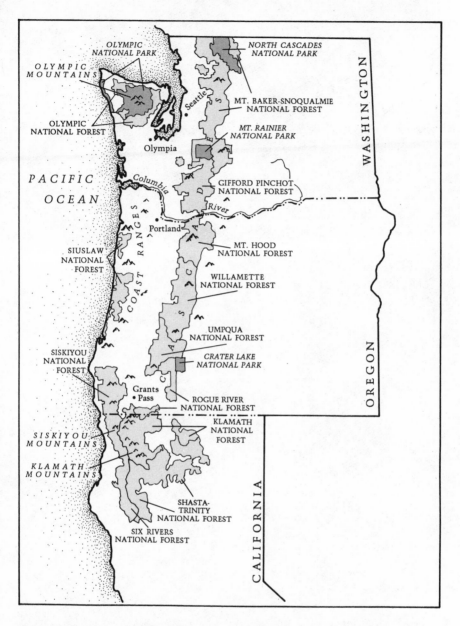

The Westside: western Washington, western Oregon, and northwestern California. *A century ago, most of this region was covered by lush ancient forest dominated by giant Douglas-firs and Sitka spruces. Now, nearly all remaining ancient forest is located on federal lands—national forests, national parks, and land managed by the Bureau of Land Management—in the Olympic Mountains, the Coast ranges, the Siskiyou and Klamath mountains, and in the Cascade Range.*

I
Ancient Forests: Global Resource, Global Concern

The Problem

Rio de Janeiro, March 8—Brazil and seven other South American nations that form the Amazon Pact today denounced "foreign meddling" on the issue of preserving the rain forest they share.

The pact nations . . . threw their full support behind Brazil, which has been accused by environmentalists and industrialized countries of failing to protect the world's largest rain forest. . . .

. . . the general secretary of Brazil's Foreign Ministry, Paulo Tarso Flecha de Lima, described the environmentalist accusations as part of "a campaign to impede exploitation of natural resources in order to block [Brazil] from becoming a world power."

"The developed countries are not the most prodigious examples when it comes to the environment," he added.

(Mac Margolis, "Amazon Nations Back Brazil on Rain Forest," *Washington Post*, March 9, 1989)

Picture the following: At the edge of the Asian, African, or Latin American rainforest, the birds have stopped singing, the mammals have fled as men armed with chainsaws fell the giant trees. Then they burn the land and plant intensively managed crops that cannot support sensitive

5

forest species. In short order, a highly complex ecosystem, whose interacting parts had survived and evolved through eons of change, is gone.
Americans have heard a lot about tropical deforestation. We have
learned that forest ecosystems moderate climate, create soils, protect
water supplies, break down pollutants, generate new medicines, and
provide homes for millions of kinds of living things. We have seen that
cutting ancient forests benefits some people, but that the costs are long-
lasting, often permanent, and are paid by everyone. But many of us do
not realize that the destruction of ancient forests is not confined to
desperately poor tropical countries. Precisely the same thing is happening to the ancient forests of our own lush, green Pacific Northwest.

As Haiti, El Salvador, and Ethiopia have, our government is now
eliminating the last sizable tracts of lowland virgin (uncut) forest in the
contiguous United States: the ancient (old-growth) forests of western
Washington, western Oregon, and northwestern California.

It is not hard to understand why ancient forests are cut. The giant
trees are among the world's finest sources of timber. And although the
timber market is plagued by sharp fluctuations, when prices are high,
the timber industry brings hundreds of millions of dollars annually into
the Northwest's economy. It provides jobs and a way of life for more than
100,000 workers.

Just as cotton shaped the environment, economy, sociology, and politics of the Deep South, timber shaped them in the Northwest. So
important was the cotton culture to planters, mill owners, and workers
that they fought the deadliest, most divisive war in our history to
preserve it, and our nation still suffers from its effects more than a
century later. But in the end, the cotton culture disappeared, a victim of
new technology, substitute products, ecologically unsound land management, competition from other regions, and overwhelming political
opposition to its practices.

The same problems now face the Northwest's timber industry. Like
the cotton industry during its decline, the timber industry is fighting to
maintain the old way of doing business. Its political influence is still
enormous, but it is facing a mounting wave of public concern for the
future of our ancient forests.

And for good reason. These crown jewels of America's forests are
being destroyed and fragmented much faster than previously thought.
About 87 percent are already gone, a loss far greater than that of the
wetlands and tropical rainforests whose destruction has garnered far
more attention. At current rates of logging, all unprotected ancient

Ira Spring

Ancient Douglas-fir/western hemlock forest, Mt. Baker-Snoqualmie National Forest (Washington). *The ancient forests of the Pacific Northwest have more species of giant conifers than anywhere else on Earth. The tallest recorded Douglas-fir reached 385 feet, but today very few reach 300 feet. Nearly all remaining ancient forest is on federal lands.*

forest in western Washington and Oregon will be gone by the year 2023. The last stands in Olympic, Gifford Pinchot, and Siskiyou national forests will be gone by 2008 and could be irreparably fragmented by the early 1990s.

For sensitive species such as spotted owls, the fragments that remain uncut will likely be too small and isolated. Genetically distinct populations and species will face extinction.

One reason for this situation is that the scientific study of ancient forests was scant until the 1970s. Before then, what little study of ancient forests there was had just one objective: to find the best way to log them. Now scientists see ancient forests in a new light. We have come to understand that they are more than just timber, more than just trees: They are ecosystems, complex systems of living things whose interactions are intricate in ways that nonliving systems are not. Their values—not just as wood and pulp, but as homes for owls and elk, as providers of services essential to people in the Northwest, the nation, and the world—are just beginning to be understood.

Ancient forests provide gourmet foods and promising treatments for disease. They build rich soils and prevent their loss through erosion. They cleanse pollutants from the air. They forestall greenhouse warming by storing more carbon than any other terrestrial ecosystem. They prevent floods and provide clean water for young salmon and municipal water supplies. They harbor the genetic diversity needed to sustain timber production in a changing world.

No less important, ancient forests have a transcendent aesthetic and religious value in the inner landscapes of natives and newcomers alike. Their majesty inspires comparison with the great cathedrals. Their haunting beauty and solace attract growing numbers of Northwesterners and visitors who seek connection with a wild world that is everywhere gone or going fast. Ancient forests are a national and international resource of the highest value.

Ancient forests are disappearing because our friends, neighbors, customers, and constituents are logging them. Timber built the Northwest, and the timber industry has almost always had its way. A century ago it acquired the best forests and liquidated them. Most old-growth that remains is on lands managed by the Forest Service and the Bureau of Land Management (BLM), ostensibly to benefit all Americans. Unfortunately, although federal laws require these agencies to balance various uses, their actions show that they manage our forests mainly to benefit the timber industry.

Agency officials point out that Congress (under pressure from the industry) requires the sale of excessive amounts of timber. To hasten the removal of our remaining old-growth, Congress has lavishly funded an environmentally damaging system of logging roads. Sometimes reluctantly, but more often not, the agencies that comply with Congress's directives are "mining" the ancient forests, destroying them forever, rather than managing them as a renewable resource. Although something resembling ancient forests (to the untrained eye) might replace them, logging destroys them forever because the management agencies will not allow forestlands the many centuries needed to produce old-growth once more. Ancient forests are being cut as if there is no tomorrow.

Not surprisingly, their fate has generated intense controversy, and, as in most controversies, truth has often been a casualty. Some environmentalists have ignored the economic concerns of the timber industry that must be heard. Some have exaggerated numbers of species that depend on old-growth before scientific evidence can verify these estimates or have overused words such as "fragile" and "unique." This hyperbole is unnecessary and counterproductive. The reality is that ancient forests provide so many benefits to so many people besides fallers and mill workers—from deep spiritual values to promising anticancer medicines to growing nontimber economic benefits—that there is an increasingly powerful political argument for their conservation.

On the other side, the timber industry has never hesitated to use specious arguments to avoid its inevitable weaning from logging ancient forests. Some elements of the industry have portrayed the old-growth question as a simple choice between wasting decadent trees to please a handful of environmental radicals versus providing lumber, paper, and jobs for the good of all. The industry has disseminated blatant falsehoods about the benefits of logging to wildlife. It has painted an overly optimistic picture of its future, given that it depends so heavily on a fast-disappearing resource that is not being replaced. It has blamed mill closures on environmentalists to divert attention from its own management failures. And it has not justified the need to cut the last publicly owned old-growth when it could have provided an ample supply of second-growth on lands it cut decades ago. The industry's influence has long protected it, but now, as the imminent end of the old-growth approaches, it is time to reappraise the cost of its privileged position.

The dispute over the fate of ancient forests contrasts different values and views of the future. At one end are people who clamor for the fastest possible conversion of ancient forests to cash. At the other end are people who feel that ancient forests have intrinsic value that cannot adequately be expressed in dollars. In between are those who seek ways to maintain both timber production and the special benefits of ancient forests on a sustainable basis. But many Northwesterners do not fit the stereotypes. Not everyone in the industry favors liquidating the last old-growth; most environmentalists recognize the role of timber in the Northwest's changing economy, and the federal officials who make the decisions are not always in the middle.

The Northwest has a unique opportunity to preserve its diverse biota and forest ecosystems while maintaining a steady flow of timber. Most developed nations have forever lost the chance to maintain viable portions of their original forests. And Third World nations face such serious economic pressures and shortages of trained manpower that prospects for sustaining their ancient forests are slim.

In contrast, the Northwest still has some intact old-growth, and our country has sound laws that can balance the interests of various groups, present and future ones. At present, however, there is little semblance of balance: On lands owned by the American people, the timber industry is logging at a record rate, ancient forests are fast disappearing, and yet, ironically enough, industry jobs are declining.

To ensure that Americans derive maximum benefits from our ancient forests, the major players—timber companies, other forest users, conservationists, states, counties, large cities, local communities, Congress, the agencies that manage public lands—need to reexamine the values of public forestlands and our options for maintaining them.

Our decision-makers especially need to reexamine the tacit, all but universal assumption that all other values are subordinate to timber production. The National Forest Management Act of 1976 clearly states that timber is only one of the important products of our national forests, yet the current practice of equating timber management with forest management will eliminate all but 6 percent of the ancient forests in our lifetimes. Further, rushing to cut everything now not only guarantees loss of their extraordinary environmental values; it guarantees harm to the timber industry in the future as well.

The fate of the ancient forests is not just a local or regional issue, one in which only Northwesterners are affected. In a world where developments abroad increasingly influence what happens here, it is a national and global issue for at least four reasons.

First, the future of the Pacific Northwest is tied to the Pacific Rim. The Pacific Rim economy will continue to expand international trade and attract people who value the Northwest's combination of vibrant economy and superb quality of life. Economic growth will depend substantially on the region's ability to sustain amenities—the close proximity to nature, the beautiful vistas, the extraordinary air and water quality, the diverse, high-quality recreational opportunities—that give it an advantage over competing regions.

Second, the health of Northwest forests is intimately linked to climate, but climate is changing. Climatic change will affect coastal human and ecological communities, agriculture, tree plantations, and natural forests. And these forests play a unique role in preventing climatic change. Changing climate jeopardizes so many interests, especially those dependent on long-lived resources such as trees, that it could actually create a climate of cooperation between previously conflicting interests: the timber industry and environmentalists.

Third, how we manage our forests sets a precedent for other nations. The United States is a world leader in conservation, and our government encourages other nations to conserve their forests in the face of pressures to liquidate them. Just as other countries look to our achievements in science, technology, economics, and culture, they watch carefully how we treat our natural resources. If our nation wants Brazil, Colombia, Ecuador, Peru, and Bolivia to save the ancient forests of the Amazon basin, we must demonstrate the will to conserve our own ancient forests. Prospects for convincing others depend on the example we set in our own Pacific Northwest.

And fourth, lest we forget, the national forests and BLM lands are owned by all the American people, not just the timber industry or the residents of timber-dependent communities. They need to be managed as the extraordinarily valuable resources they are, to benefit all Americans now and forever.

At a time of rapid change and uncertainty, our chances of having an economically vital, livable Pacific Northwest depend on *preserving our options*. Because Northwestern forests recover from disturbances so slowly compared with crops, livestock, fisheries, or game populations, the effects of today's decisions will be visible for many lifetimes. It is foolish to fritter away our options in hope that someone might be able to fix things someday. A more prudent way to provide enough timber, water, fisheries, wildlife, and recreation is to develop a sustainable strategy for preserving and managing our forestlands.

Sustainable forestry means more than ensuring a steady timber flow. It means protecting our forests and keeping them healthy to provide options for meeting our future needs. All of us—rangers and senate staffers, lawyers and citizen activists, loggers and backpackers—are shaping the world for generations to come. Undoubtedly they will judge us on how well we preserved their options.

The Focus

... these forests of trees—so enchain the sense of the grand and so enchant the sense of the beautiful that I linger on the theme and am loth to depart ... such forests, containing firs, cedars, pine, spruce and hemlock, envelop Puget Sound and cover a large part of Washington Territory, surpassing the woods of all the rest of the globe in size, quantity and quality of the timber.
(Samuel Wilkeson, *Notes on Puget Sound* [1869])

This book focuses on the forests of giant conifers of western Washington, western Oregon, and northwestern California, the region west of the crest of the Cascade Range. Foresters often call it the Douglas-fir region after its most important source of timber, but some of its forests are not dominated by Douglas-firs, and Douglas-firs also dominate some forests to the east and south. The term "Pacific Northwest" is not perfect either. The forests of Idaho, eastern Washington, and eastern Oregon are quite different; political boundaries seldom coincide with biogeographic ones. For lack of a better term, I am adopting the one used in Washington and Oregon: the "Westside."

Within the Westside, most lower elevation forest is dominated by Douglas-fir except for a coastal strip that is dominated by Sitka spruce. These forests are the subject of this book.

Similar Douglas-fir forests range northward into southern British Columbia, and similar Sitka spruce forests extend northward along the coast to southeast Alaska. From extreme southwest Oregon southward, coast redwoods dominate a thin coastal band. At higher elevations, Douglas-firs are replaced by firs (true firs belong to the genus *Abies*, while Douglas-firs belong to *Pseudotsuga*) and mountain hemlocks. And east of the Cascades are forests of Douglas-firs, ponderosa pines, firs, and other species (see the essay by Peter Morrison below). The similar ancient forest ecosystems of British Columbia and the various ancient forests to the south, east, and at higher elevations all deserve their own treatment—they are important and they, too, are disappearing—but time and space do not permit it in this book. Nonetheless, both their similarities and differences provide some useful comparisons with lowland Westside forests.

Ancient Forests East of the Cascades

□ *Peter H. Morrison, The Wilderness Society, Seattle, Wash.*

Although most of the Northwest's ancient forests lie west of the Cascade crest, there are still areas to the east with magnificent old-growth stands, even some that rival the finest on the Westside. Ancient forests on the Eastside—the region of Washington, Oregon, and California east of the crest of the Cascades—range from dry ponderosa pine stands to wet western redcedar stands nearly identical to those in coastal forests. The Eastside is a more extreme environment, both hotter and colder, and mostly very dry, especially on south-facing slopes. But sites in valley bottoms can be moist and productive. The diversity of forest ecosystems reflects this environmental complexity.

Some particularly notable areas are in Wenatchee National Forest, where deep tephra (pumice and ash) deposits from Cascades volcanoes form highly productive, moisture-retaining soils in valley bottoms. Here ancient Douglas-fir frequently reach 32 inches in diameter and over 200 feet in height, with exceptional ones over 6 feet in diameter and nearly 300 feet in height. In one such stand I discovered a gnome-plant, a rare saprophytic member of the heath family. The closest previous record of this species was from an ancient forest on the Olympic Peninsula. Elsewhere on wetter sites in this zone are groves of western redcedars reaching over 10 feet in diameter. One valley with stands of giant western hemlocks, western redcedars, and Douglas-firs is now threatened with logging and by inundation by the Bumping Lake Reservoir. Another with more extensive stands was cut and flooded years ago.

Elsewhere are drier forests where Douglas-firs are both the dominant seral and climax tree. Other ancient forests have Douglas-fir and ponderosa pine in a parklike landscape with a lush understory of pine grass.

The driest ancient forests consist of open ponderosa pine stands that extend like fingers into shrub-steppe country.

Even in these open stands, pines attain immense size, towering 200 feet and reaching diameters in excess of 8 feet. On one recently blown-down ancient ponderosa pine, I walked about 60 feet up the trunk to the first lateral branch, which had reached a diameter of over 3 feet.

There have been few studies on the ecological characteristics of Eastside ancient forests, and ecologically based old-growth definitions are still in a formulative stage. But, in general, Eastside and Westside old-growth forests have the same characteristics: large live trees, multi-layered canopies, large snags, and logs. Snags and downed logs are less abundant in the Eastside because the fire frequency is higher and fuels are drier, so that they decay and burn more readily. Diameters of old trees are not as great because growing conditions are harsher. Stands on drier sites tend to be more open. In some cases, old-growth stands may be composed of a single species, with a wide diameter distribution of large and small trees.

Little old-growth remains on the Eastside. Its original extent was limited, but Eastside ancient forests have been hit particularly hard by logging and continue to be cut at a very high rate. Their rapid loss is due in part to their location in accessible valley bottoms. On many Eastside national forests, old-growth stands may be the only ones that can be logged at a profit.

Until recently, these forests have not received the public attention of Westside forests because they are farther from major population centers. This is regrettable because the most imminently endangered Pacific Northwest ancient forests are east of the Cascades. □

A Note to the Reader

... the most difficult feat on earth: transferring one thought from one mind to another.
(James Lipton, *An Exaltation of Larks, or the Venereal Game* [1977])

My aim in this book is to weave the strands of scientific understanding of ancient forests together in a way that makes sense for thoughtful laypeople, yet has the rigor required by professionals. This is not an easy task. One reason is that these ecosystems operate on a timescale difficult to comprehend; humankind has gone from the Wright Brothers' plane to the space shuttle in just 7 percent of a Douglas-fir's life span. Another is that ancient forests are complex; appreciation of their complexity is essential to replace the simplistic notions that have hastened their destruction. Moreover, forest processes are inherently cyclical while books are inherently linear; I found myself wanting to put each chapter ahead of the next. But because each chapter builds on previous ones, unless you are interested in only one specific topic (and not the way it relates to the forest as a whole), it is probably best to read the book from beginning to end.

To make the reading comfortable for nonscientists, I have not provided a citation for every statement that would be cited in a publication for scientists. Rather, I refer to key works by giving the authors' names and the dates of their publications. All these and other useful publications are listed by chapter in the Suggested Readings at the end of the book.

Any decision on how to present measured units is a good way to start a fight. Because many forest managers and forestry researchers use the English system (which the English are using less and less), but nearly all ecological researchers use the metric system, this book uses each as appropriate. Often I give both measures at the beginning of a discussion to provide some context. This, or a conversion chart, allows readers to make any necessary conversions. After their first mention in a section, measures are sometimes abbreviated, such as "a." for acre. Similarly, I use a few chemical symbols such as N for nitrogen or CO_2 for carbon dioxide where appropriate.

Throughout the text, when I cite the Forest Service, I mean the agency of the U.S. Department of Agriculture. As noted before, BLM stands for the Bureau of Land Management, an agency of the U.S. Department of the Interior.

In Chapter 2 I briefly sketch the forests, their geographic setting (emphasizing the importance of climate), and the history of their use by humans.

In Chapter 3 I discuss three keys to understanding the ancient forest conflict: (1) the central role of ecological succession; (2) how what seems to be a forest waste product, namely, dead wood, is actually vital to a healthy forest; and (3) a precise, scientifically meaningful definition of ancient forests.

The fourth chapter introduces the concept of biological diversity and discusses the most familiar kind of biological diversity, the diversity of species, starting with the spotted owl and ending with one of the most important but least appreciated groups of ancient forest dwellers: the fungi.

Chapter 5 looks at the values of less familiar levels of biological diversity, the one below species diversity—namely, genetic diversity—and the one above it—namely, ecosystem diversity. Ancient forests are extraordinary not only as habitat for species, but also as a genetic library for forestry and as providers of essential, free services. The chapter also discusses the scientific importance of ancient forests and how these forests differ from the tree plantations that are replacing them.

Chapter 6 examines one of the major threats to ancient forests: timber operations, from roading to fires suppression, separately and then in combination.

The next chapter analyzes the unprecedented but little-appreciated external threats to Westside forests that result from atmospheric pollution; in combination, timber operations and climatic change jeopardize forests that might survive either. This chapter suggests that loggers and environmentalists should be worrying about something besides one another: changing climate.

In Chapter 8 I discuss sustainable forestry as an alternative to the destruction of ancient forests, starting with a current assessment of how much old-growth remains and ending with a look at the need for a mature environmental ethic to resolve the ancient forest controversy.

In Chapter 9 I offer conclusions and recommendations, which, I believe, put the issue in its global perspective and provide a foundation for sustainable forestry in the Pacific Northwest by conserving the basis of timber production: biological diversity.

Finally, there is a glossary of scientific and forestry terms, a bibliography of suggested readings for each chapter, and an index including scientific names used in the text.

Only time will tell whether we can find the wisdom and courage needed to get the fullest benefits from our forests. Previous generations did not have our scientific knowledge or appreciate the Earth's finiteness. They eliminated the ancient forests of Massachusetts, North Carolina, Michigan, and Texas, and we—all of us—are forever poorer. Whether our generation will eliminate the last ancient forests is uncertain. But one thing is certain: Time is running out.

II

The Forests of the Pacific Northwest

Biological Uniqueness

THE LAST GREAT VIRGIN FORESTS

[I] never beheld so fair a thing: trees all along the river, beautiful and green, and different from ours, with flowers and fruits each according to their kind, many birds and little birds which sing very sweetly. (Christopher Columbus, *Journal of the First Voyage* [1492])

When Europeans "discovered" the New World, they were stunned by its richness. In the Old World, the vast herds of mammals were long gone. The remnants of once-bountiful game were reserved for the rich and the royal. The forests of Italy and Spain had been cut over and grazed for millennia. The forests of England and the Netherlands were largely a memory when the early colonists embarked in their wooden ships. In the densely peopled continent they left, nature had long been subdued.

But in America, a dense cloak of virgin forest covered an endless land. There were no old dirty cities, no powerful nations, only a scattering of natives who lacked modern weapons and resistance to European diseases. There were fish to catch, game to shoot, deep soils that had never felt the plow, and huge trees to cut for stockades and houses, ships and fuel. The virgin forests were the primary resources for the new Americans building a new nation.

As the tide of pioneers flowed west, virgin forests yielded to the axe. In

17

two centuries, the pioneers reached a land where forests gave way to prairies. In their haste, they missed some stands, especially on slopes too steep to cut, but, in time, except for a smattering of tiny patches, the virgin eastern forests were gone. Gone with them were eastern elk, bison, timber wolves, cougars, Carolina parakeets, ivory-billed woodpeckers, and passenger pigeons. With their habitat gone, they were easily overhunted.

But the world's richest virgin forests lay waiting west of the prairies and mountains. Soon they too began falling. But unlike the East, more than a century later some still remain. This makes the Northwest unique: No other region of the lower forty-eight states with trees suitable for timber still has substantial tracts of uncut forest. The Pacific Northwest still does, and remarkable forests they are.

THE UNIQUE CLIMATE

Prevailing winds vault over the Olympics, over the Cascades, over the Rockies. Each time they leave most of their moisture on the western side of the range and little remains to water the land lying on the lee side to the east. Oregon and Washington are lush and green west of the Cascades, so dry they are often almost deserts east of the mountains. (Edwin Way Teale, Autumn Across America: A Naturalist's Record of a 20,000-Mile Journey Through the North American Autumn [1956])

Because trees can grow only under certain conditions, a large portion of the Earth's surface is treeless. Dense forests occur only where geographic factors occur in the right combination. Until about 18,000 years ago, the Westside was either glaciated or covered by treeless tundra vegetation, indicative of cold, dry conditions. Today conditions are very different. A nearly ideal coincidence of climatic factors between the crest of the Cascades and the Pacific Ocean has yielded what Jerry Franklin and C. T. Dyrness call "the classic coniferous forests of the world" (1973, p. 53).

The Cascade Mountains that run from British Columbia to northern California set the Westside apart from the rest of the country. Although their crests lie only 30–130 miles from the Pacific or Puget Sound, many Cascade peaks exceed 5,000 feet, with some volcanoes attaining far greater heights, including Mt. Rainier (14,410 ft.) in Washington, Mt. Hood (11,235 ft.) in Oregon, and Mt. Shasta (14,162 ft.) in California. Some have been dormant for thousands of years; others have been active recently. Washington's Mt. Saint Helens experienced a cataclysmic eruption in 1980. Lassen Peak in California erupted as recently as 1917. Mt. Rainier was last active about 160 years ago.

Paralleling the Cascades along the coast are several mountain ranges. The highest are the Olympics in northwest Washington (maximum elevation: 7,965 ft.) and the somewhat lower Siskiyou and Klamath mountains in southwest Oregon and northwest California. Between them are smaller mountains collectively called the Coast Ranges. Between the coastal mountains and the Cascades lie the Puget Lowlands in Washington and the Willamette, Rogue, and Umpqua valleys in Oregon.

Together, the Pacific Ocean and the mountains profoundly influence the region's climate. As winds from the Pacific ascend the slopes, the moisture-laden air cools and produces frequent rains, making the Westside the wettest part of the contiguous forty-eight states. The Cascades are so effective in precipitating moisture that the Eastside region is an arid "rain shadow." For example, while annual precipitation of 70–100 inches is common on the Westside, Spokane, Washington, and Pendleton, Oregon, on the Eastside receive only seventeen and twelve inches, respectively, yearly. The coastal mountains also produce their own rain shadows, albeit less intense ones. Quinault, Washington, on the west side of the Olympic Peninsula, averages 133 inches annually, while Seattle, in the rain shadow, gets only about 35 inches, less than New York City or Atlanta. Fog is common, especially in coastal areas and some valleys perpendicular to the coast. In general, west-facing slopes of the coastal mountains are the wettest, the west slopes of the Cascades are a bit drier, and the lowlands between them are drier still.

The Westside's temperature regime is exceptionally mild and has been for about 6,000 years, as suggested by fossil pollen records analyzed by Linda Brubaker (forthcoming) of the University of Washington. Coastal Washington has fewer freezing January nights than northern Louisiana, yet in July daily maximums are lower than in northern Minnesota or Maine. The frost-free season is long, and skies are cloudy for most of the year. There are no hurricanes and virtually no tornadoes, although some winter storms generate hurricane-force winds.

In the northern hemisphere, only western Europe occupies similar latitudes on the west side of a continent and has temperate maritime conditions. But the Coast Ranges and Cascades that wring moisture from the prevailing winds make the Westside far wetter. Still another factor makes its climate unique. Although renowned for wetness, it experiences summer drought unknown in moist regions of east Asia, eastern North America, or western Europe.

Within the Westside, the importance of drought varies both north-south and east-west. The north-south gradient results from movements of gigantic Pacific High and Aleutian Low pressure cells. The Pacific High is a fair weather system between Hawaii and North America. It moves northward in late spring and early summer and returns southward in fall and early winter. The Aleutian Low generates storms that thrust southward from the Gulf of Alaska to batter the Northwest, but in summer the Pacific High tends to block them, keeping rainfall well to the north. The seasonal dance of these weather systems causes the Westside's unique alteration between dry summers and wet winters.

In southeast Alaska, rain falls year-round, but the length of the rainy season decreases to the south. Most of the Pacific Northwest receives less than 10 percent of total precipitation during the summer months. The dry season often lasts two to three months in western Washington, but in southwest Oregon and northwest California, it approaches six months. Coastal areas are more likely to catch an occasional summer rain than those in the rain shadows.

Westside winters are mild. At low elevations, hard freezes are rare and temperatures tend to hover just above the freezing point for long periods. Northern areas and higher elevations are cooler, and a large fraction of precipitation falls as wet snow on the upper slopes of the Cascades and the Olympics. Snow is less common in the Coast Ranges, the Siskiyous, and the Klamaths, but even they can receive substantial snows in some years.

Because plants are sensitive to variations in weather, soils, and fire frequency (all of which are shaped by climate), the ecological communities of the Pacific Northwest mirror geographic variation in climates. As a result, they are also likely to be sensitive to changes in climate.

THE GIANT CONIFERS

This is the forest primeval.
The murmuring pines and the hemlocks . . .
Stand like druids of old.
(Henry Wadsworth Longfellow, *Evangeline* [1847])

Seventy, 80, even 100 million years ago, when dinosaurs roamed the Earth, the world's forests were dominated by conifers. Indeed, some paleontologists believe that the long necks of *Brachiosaurus* and its relatives were adaptations for feeding high in coniferous canopies. Flowering plants (angiosperms) were new then, mostly tiny things. But today, flowering plants display dazzling diversity, with 250,000 species from supple grasses to vibrant orchids, from tiny duckweeds to massive oaks. They dominate most terrestrial ecosystems.

In contrast, the 700 or so species of gymnosperms (mainly ones we call conifers) are all woody plants. Most are relegated to colonizing lands eventually dominated by broadleaf trees or occur in places too soggy, sandy, dry, or frigid for the angiosperms that exclude them from the best habitats. In a languid race lasting eons, conifers have fallen behind flowering plants. They have lost their former magnificence almost everywhere, except the Pacific Northwest.

To one accustomed to forests anywhere else, Westside forests are so big, lush, and diverse that it is difficult not to use superlatives. The trees are exceptionally long-lived. The species that dominates much of the region often attains 500 years and can reach 1,200; the longest-lived species can reach 3,500 years. These forests maintain high productivity over exceptionally long periods, but they also produce unparalleled amounts of huge dead standing trees and fallen logs that decompose slowly. As a result, they support the largest or second largest accumulations of living matter anywhere, have the world's largest accumulations of dead trees, and store more carbon than any other terrestrial ecosystem thus far measured. For their latitude, they are unusually rich in wildlife species. And they undergird some of the world's richest fisheries by providing spawning grounds for Pacific salmon.

But of all their features, the one that immediately grabs a newcomer is their titanic scale: Uniquely dominated by a diversity of huge conifers, they dwarf forests almost anywhere else. The largest tree in eastern forests (Eastern white pine) once reached 200 feet at most. But at least thirteen species of Westside conifers reach 200 feet, the most anywhere on Earth, and the tallest approaches 400 feet.

Two of the most magnificent conifers actually fall just outside the scope of this book. Giant sequoias, the world's most massive living things, are now confined to tiny, scattered stands in the Sierra Nevada in California. Coast redwoods, the world's second tallest trees (some approached 400 feet, and the largest now exceed 360 feet), occupy a thin coastal strip from extreme southwest Oregon to central California; they would be included in this book were they not already extensively treated elsewhere. Excessive logging has reduced redwood forests to a mere remnant of what they once were.

But coniferous forests reach their zenith of glory from northwestern California through western Oregon, Washington, British Columbia, and southeast Alaska. The Pacific Northwest is home to the world's largest species in every conifer genus that occurs there except for the junipers. Among these ancient giants, dinosaurs would probably feel at home.

Douglas-firs are the most valuable trees in world timber commerce, the trees that built the Northwest, the world's third tallest trees. Exceptional Douglas-firs reach 330 feet (one reportedly reached 385 feet) and attain diameters of 16 feet, dwarfing the seven or so other species of *Pseudotsuga.* (Maximum sizes for the trees in this section are from historic records; today's largest trees are almost always smaller because the biggest ones were logged first.) Douglas-firs occur in more kinds of ecosystems than any other Northwest tree and dominate forests from California to British Columbia. While most giant conifers begin life in serene shadow, Douglas-firs are conceived by fire and grow best in sunlit openings. And on the best sites they grow rapidly; one attained 170 feet in seventy-two years. The strong wood of old trees makes superb framing lumber and plywood.

Western redcedars occur throughout moister parts of the Pacific Northwest, typically scattered among other conifers but occasionally forming pure stands in low spots where the roots of other conifers would drown. This shade-tolerant species is the largest of the world's six or so *Thuja* spp., reaching 200 feet tall and 20 feet in diameter, yet it is fairly slow-growing. Once the most important tree to the Westside's Native Americans, large redcedars bring the highest prices of any widespread Northwest species because their easily split, decay-resistant wood is unsurpassed for making shingles.

Western hemlock is the largest of the world's eighteen or so hemlocks. This moisture-loving species is probably the most abundant conifer from coastal Oregon to southeast Alaska. Hemlocks reach up to 215 feet tall but attain 10-foot diameters at the very most. Faster-growing than redcedars and far more prolific, their delicately beautiful seedlings often blanket old nurse logs on the forest floor. Hemlocks are very shade-tolerant, and individuals that persist long enough beneath the canopy can form extensive pure stands when dominant trees succumb. They are cut extensively for lumber and paper pulp.

Sitka spruce is by far the tallest of the world's fifty or so spruces. It towers over the redcedars and hemlocks in many foggy coastal forests from California to southeast Alaska that are too wet to favor Douglas-fir. The second tallest Westside tree, exceptional spruce reach 300 feet and 17 feet in diameter and can add a foot to their girth per decade. Unlike most fast-growing trees, spruce can reproduce in forest shade; like western hemlocks, many get their start on fallen nurse logs. Their light wood is one of the strongest in the world for its weight and was essential for building World War I fighter planes.

Sugar pine reaches heights to 250 feet and diameters to 18 feet, the giant among the world's 120 or so pines. It is mainly a California Sierran species that extends into drier forested areas of southern Oregon. Like Douglas-firs, sugar pines are not very tolerant of shading and cannot readily replace themselves under closed canopies. Named for their sweet sap, they have the world's longest pine cones (some to 23 inches) and make valuable lumber.

Ponderosa pine is a very widespread western tree that can attain heights to 232 feet and 8 feet in diameter. It is absent from most Westside forests but occurs in the Willamette Valley and in the hotter, drier areas of southwest Oregon and northwest California. It is deep-rooted and well adapted to dry conditions. Old trees are prized for their lumber.

Western white pine is another giant pine, with exceptional individuals attaining 239 feet and 7 feet in diameter. It is more common in inland British Columbia, northeastern Washington, and northern Idaho, but scattered trees occur throughout the Cascades and Olympics. Western white pines are poor competitors on good soils but can dominate forests on gravelly or boggy Westside sites. Their lumber is exceptionally valuable.

Noble fir, reaching up to 260 feet and 9 feet in diameter, is the tallest of the world's fifty-five or so true firs. It occurs mainly in higher elevations of the Washington and Oregon Cascades, although some can be found in scattered pockets of the Coast Ranges into northwest California, where it intergrades with the very closely related Shasta red fir. Unlike most firs, noble fir fares poorly when shaded but grows rapidly when conditions are suitable. Its wood is light and strong, like that of Sitka spruce.

Pacific silver fir, which reaches 245 feet and 8 feet in diameter, is found in scattered pockets in northwest California but is far more common in the Cascade and Olympic mountains and northward into British Columbia and southeast Alaska. It occurs in the higher elevation forests inhabited by Douglas-firs, hemlocks, and redcedars and in still higher areas that accumulate deep snow. Shade-tolerant and relatively slow-growing, silver firs are cut for lumber and plywood but are an especially valuable source of paper pulp.

Grand fir attains 250 feet tall and 5 feet in diameter and can grow very quickly. In the Westside, it ranges from Vancouver Island to northwest California but is seldom abundant, occurring most commonly in low-land ecosystems drier than those favoring redcedars and hemlocks. Grand firs are used for paper pulp.

White fir can reach up to 230 feet in height and 6 feet in diameter. It extends northward from the California Sierras and occurs in northwest California and southwest Oregon at higher elevations with Douglas-firs and sugar pines. It is highly shade-tolerant and is used for lumber and paper pulp.

Incense cedar is the largest of the three *Calocedrus* spp., reaching up to 225 feet and 12 feet in diameter. An inhabitant of sites much drier than those having western redcedars, it occurs in the southern Oregon Cascades, Siskiyous, and Klamaths south into the Sierra Nevada. Its fragrant, rot-resistant wood is used for making cedar chests, fence posts, and billions and billions of pencils.

Port Orford cedar, found only within a small area along the southwest Oregon and northwest California coasts, reaches a maximum of 240 feet and 11 feet in diameter, the largest of the eight or so species in its genus. Its wood—light, strong, rot-resistant, shrinking and swelling little when dried or soaked—is used to make arrow shafts in the United States and as top-quality construction lumber in Japan. A large Port Orford cedar can bring a higher price than any other Westside conifer.

CLIMATE AND THE DOMINANCE OF CONIFERS

The thirsty earth soaks up the rain,
And drinks, and gapes for drink again.
The plants suck in the earth, and are
With constant drinking fresh and fair.
(Abraham Cowley, *Anacreon* [1656])

The giant conifers and a diversity of smaller pines, firs, hemlocks, spruces, cedars, and yews form the world's finest coniferous forests. But what is it about the Westside that favors conifers over the broadleaf trees prevalent in most temperate regions?

The likeliest answer reflects the different ways that the region's unique climate affects conifers and broadleaves. As Dick Waring and Jerry Franklin (1979) explain, in most of the world's temperate forested areas, precipitation occurs year-round or mainly in the warm months. Trees receive moisture when warm temperatures and high light levels allow trees to photosynthesize rapidly. Where summer moisture is adequate, broadleaves usually outcompete conifers.

But in the Pacific Northwest, summer temperatures and light levels, which create higher respiratory and evaporative demands, occur just when moisture is least available. In some tropical forests, broadleaf trees lose their leaves during the dry season and regrow them in the wet season. But most broadleaves in Westside forests cannot do that because they have another evolutionary Achilles' heel: At subfreezing temperatures they cannot photosynthesize. Wherever such temperatures occur (including most of the Westside), most broadleaf trees are deciduous, dropping their leaves as winter approaches.

Thus, broadleaves are caught in a double bind. They have difficulty maintaining leaves through both Westside summers and winters, and leaflessness does not promote high growth rates! Where soils are moist during the summer (e.g., on floodplains), deciduous angiosperms, such as red alders, black cottonwoods, Oregon ash, and bigleaf and vine maples, do very well. And in the southern part of the region, where winter temperatures seldom dip below freezing, drought-resistant, thick-leaved evergreen angiosperms, such as madrone, tanoaks, oaks, and golden chinquapins, are far more abundant than farther north.

But on sites where soils dry during summer and where freezes occur regularly during winter—most of the Westside—conifers predominate. Conifers are more tolerant of summer drought than deciduous broadleaves and use the abundant precipitation in the cool months to photosynthesize when angiosperms are "shut down."

Indeed, the Westside climate uniquely favors conifers. Nowhere else in the northern hemisphere has the peculiar combination of ample rain, summer drought, and occasional subfreezing temperatures. For example, winter weather in the British Isles is similar, but there is no summer drought. Hence, deciduous broadleaves naturally dominate the forests. Only when foresters discourage native broadleaves do Westside Sitka spruce and Douglas-fir grow well in the British Isles.

Westside trees are sensitive to climate. Seemingly small climatic variations, such as those caused by topography, play a major role in shaping tree distributions within the region. Sitka spruces are confined to the wettest, foggiest coastal areas and to some river valleys where rainfall is high, summer fogs are frequent, and fire frequencies are low. Western redcedars and western hemlocks do especially well there too, but they also occur in drier lowlands and on the slopes of the Cascades, where Douglas-firs dominate.

More conifer species occur to the south, where the diversity of climatic conditions creates a zone of overlap between Pacific Northwest trees such as redcedars and Sierran trees such as incense cedars. In the southern Oregon Cascades, the Siskiyous, and the nearby Klamath Mountains in California, the slopes with the coolest aspects—those facing north and northeast—are southerly outposts of species that cannot tolerate hot, droughty conditions, such as Alaska-cedars and western hemlocks. The warmest slopes—those facing south and southwest—are northerly outposts for the species that tolerate hot, dry summers, such as sugar pines and incense cedars. The presence of species from both cooler and warmer climates makes the Klamath and Siskiyou mountain areas of exceptionally high diversity.

Altitude is also important to Westside trees. Douglas-fir seedlings have difficulty in areas that get deep snowpacks, but silver firs are more tolerant of snows and become more abundant at higher elevations. Mountain hemlock and subalpine fir are consistently found above the Douglas-fir and Pacific silver fir zones. Most Pacific Northwest species found in Washington's lowlands tend to occur only at higher elevations to the south. In general, lowland coniferous forests have more diverse trees than higher ones.

However odd it seems for the wettest region of the contiguous United States, fire is a dominant force in much of the Westside. In general, it occurs with greater frequency in warmer and drier areas southward and inland. Only the swampiest redcedar stands and the wettest, foggiest, coastal Sitka spruce forests seem untouched by fire.

Fire affects tree species in different ways. Douglas-firs are highly dependent on fires. Because they do not germinate well on duff (decayed litter from trees) and are not very tolerant of shade, only disturbances that expose mineral soil to the light allow many Douglas-fir seedlings to survive in much of the region. Fires provide these conditions, but most other disturbances do not. Any natural Douglas-fir stand where landslides are unlikely probably began with a fire hot enough to kill the canopy trees and burn off the accumulated duff. Further, Douglas-firs (as well as Port Orford cedars and sugar and ponderosa pines) have a thick bark and can withstand cool fires.

In contrast, Sitka spruce, western redcedar, western hemlock, and firs have a thinner bark and shallower root systems, making them more vulnerable even to relatively cool fires. Further, fire does not help their establishment nearly as much as with Douglas-firs. Thus, the dominance of Douglas-firs in natural stands reflects the prevalence of fire in Pacific Northwest landscapes.

Human Impacts: A Historical Perspective

This only is denied to God: the power to undo the past.
(Agathon, ca. 448–400 B.C.E., from Aristotle's *Nicomachean Ethics* [4th century B.C.E.])

The Native Americans of the Westside were the richest north of Mexico. Unlike tribes elsewhere, these peoples inhabited an exceptionally benign land. Summer temperatures were pleasant, and winter temperatures were far warmer than those east of the Cascades. Fresh water was never in short supply, food was abundant year-round, and the forests provided limitless supplies of wood.

Many tribes inhabited the Westside, usually where forested rivers meet the sea. Their cultures were adapted to the ecology of old-growth forests, and they were more dependent on large trees than any other Native Americans. Like many others, they used wood for spears, bows and arrows, foundations, and fuel. But unlike most, they made no clay pots, using densely woven wood baskets or carefully formed bentwood boxes instead. They fashioned clothes and cordage from woven threads of inner bark fiber instead of using animal skins. Small hide or bark boats would have confined them to placid waters, so they built dugout canoes seaworthy enough to venture into the stormy Pacific in pursuit of fishes, seals, and whales. Alone among American Indians, they lived in large houses made from hewn planks.

While other Native Americans were village dwellers, who grew crops of corn, beans, and squashes, or nomadic gatherers or hunters who followed the herds of large mammals, these peoples lived in permanent settlements without agriculture. Their greatest environmental effects were probably from setting fires that favored preferred food plants, such as camas. With superabundant salmon from the rivers, bounteous finfish, shellfish, and marine mammals from the sea and diverse mushrooms, roots, berries, and game from the forest, they found agriculture unnecessary. Nowhere else on the continent could people live off the land so easily. Such abundance allowed Northwest coastal Indians the leisure needed to develop exceptionally rich cultures.

Westside Indians used some forest species more than others. Except for firewood, they left Douglas-firs largely undisturbed and made light use of western hemlock and Sitka spruce. But western redcedars (called *rpahyuhts* by tribes of the Puget Salish language group) were another matter. Old-growth redcedars were as important to Westside Indians as bison were to Great Plains Indians. Although some tribes gleaned precious mountain goat wool caught on twigs or sheared wool from their dogs, redcedar bark was the main source of fiber in much of the region. Redcedar wood was so easy to split and work that they used it for everything from arrow shafts to dugout canoes longer than a city bus. Its resistance to rotting was essential for house foundations and planking in a climate ideal for wood-decaying fungi.

Indians could not cut many redcedars. Lacking the power of draft animals and the strength of metals, their technology constrained them. Except at river mouths, their populations were sparse. Felling giant trees with fire and hand-held stone tools was an awesome task for even expert fallers. Moving logs was so difficult that Native Americans could alter the composition of forests only near their settlements. In contrast, Euro-American settlers, armed with more powerful technologies, had far greater effects on the ancient forests.

Settlers faced a daunting problem when they arrived in the Westside: clearing the land. In the East, carving fields from the virgin forest was difficult enough. But the conifers that blanketed the Westside were so much larger that clearing was an overwhelming task. Many wisely chose to settle in drier grasslands and thinly treed oak savannas of the Willamette Valley. Those who tried farming among the tall trees often found that summers were too cool for crops. The newcomers who fared best were those who cut and sold the huge trees.

The giant conifers challenged early loggers too. Although they had sharp, strong, steel axe heads that Native Americans lacked, cutting a single tree could take a pair of fallers a day or more. Use of oxen was also a major improvement, but the technology for removing logs was still primitive. But in the mid-1800s, the California gold rush dramatically

National Agricultural Library, Forest Service Photo Collection

Ancient Douglas-fir and loggers, 1918 (Oregon). *Early loggers cut the biggest and best trees first, eliminating most ancient forest in the productive river bottoms and lowlands.*

expanded the market for mine timbers and lumber for housing, stimulating competition and the loggers' ingenuity. Double-bitted axes, then two-man crosscut saws, sped the fallers' task. Flumes, then steam-powered donkeys and railroads, made removing logs much easier. Mills progressed from two-man rip saws to high-speed band saws that could produce lumber a hundred times faster.

By the early 1900s, technology no longer limited the felling of ancient trees. The only limits were economic. With superabundant trees of unsurpassed quality, negligible prices for stumpage and expanding markets, the loggers' motto was "Cut and get out." They were hampered neither by government regulation nor by any sense that the trees might someday be gone. Their only limits were the speed and efficiency in cutting and marketing logs.

Unlike the Indians, commercial loggers were not confined to using western redcedar. Their ability to mill lumber made the stronger, more abundant Douglas-fir the premier timber species. Some loggers clearcut extensive areas, leaving slash-covered lands that often burned repeatedly. Others would "highgrade" or selectively log only the biggest, defect-free Douglas-firs, spruces, and redcedars. Hemlocks brought lower prices and were usually ignored. Any trees that survived the felling and skidding of those around them were released from competition and could grow and reproduce faster. Therefore, both clearcutting and highgrading altered forest structure and species composition.

So did the location of logging "shows." Early loggers had to locate near a river or coast to minimize the costs of getting logs and lumber to market. Cutting and skidding logs were easier on moderate slopes than in the mountains. Logging in coastal lowlands and valleys also had an unintended effect: Soils are deepest and richest there, so trees grow fastest. Aided by federal land grants to railroads that the railroads then sold, timber companies amassed vast land holdings at bargain prices. Ownership of the most productive lowlands quickly passed into private hands.

The practices of the timber industry in the West spurred cries for preservation from citizen conservationists and officials in federal and state governments. In 1891, Congress gave the president authority to establish forest reserves, lands withdrawn "to preserve the forests therein from destruction." Under this authority, Presidents Benjamin Harrison and Grover Cleveland withdrew almost 40 million acres.

President Theodore Roosevelt continued the process, adding millions of acres until 1907, when Congress, under pressure from the timber industry, wrested this authority from the president in a rider to the agriculture appropriations bill. At the urging of Gifford Pinchot and others, Roosevelt protected a further 16 million acres in the week before signing the bill, to the chagrin of western lumber barons.

Roosevelt soon created what became the Forest Service to protect these lands. But as was true throughout the West, lands in federal ownership were mainly those that nobody wanted because they could not be farmed, ranched, or logged profitably. Farther from transportation, steeper, higher, snowier, with thinner soils and lower productivity than the privately owned lowlands, they became our national forests.

The sharply different productivity of Westside lands owned by the timber industry and the national forests is evident in the distribution of various site classes (Table 2.1). Foresters classify forestlands on the basis of potential annual wood production. Commercial forestlands (77 percent of Westside forestland or 61 percent of total Westside land) are those capable of producing at least 20 cubic feet of wood per acre each year and available for logging. The least productive commercial forestlands can produce 20–49 ft.^{3}/a., while the best lands are at least three to six times more productive. National forests have one-half of all Westside commercial forestland in the least productive category, with only one-fifth in the most productive category. A substantial portion of Westside national forestland does not even meet minimum criteria for the lowest productivity class.

Table 2.1

PERCENTAGE OF WESTSIDE COMMERCIAL FORESTLAND IN THREE SITE
PRODUCTIVITY CLASSES, BY OWNERSHIP

Potential Annual Wood Production (ft.^{3}/a.)	Timber Industry	National Forest	Other	Total
Poor (20–49)	23	49	28	100
Medium (50–119)	24	41	35	100
Good (120+)	42	21	37	100

SOURCE: Olson (1988).

Current ownership patterns were established by the 1920s, but for several decades, nearly all logging was on private and state lands. A few areas having old-growth were withdrawn from timber production: Mount Rainier National Park in 1899, Olympic National Park in 1938, and North Cascades National Park in 1968 (the altitude of Crater Lake National Park is too high to have the kind of ancient forest discussed in this book). Wilderness areas in which logging is prohibited have also been designated but have relatively little old-growth. Most of them are so high that, as many quip, their main crops are "rocks and ice."

The national forests were largely uncut by the 1950s, when the postwar economic expansion spurred demand for wood. Timber companies, which had liquidated most available old-growth on private lands, did not have sufficient inventories to meet surging demand. Although the national forests were far less *productive* than private forestlands—tree growth on poor sites is much slower—they had much more *standing timber* because only a small portion had been cut.

Whereas timber companies had previously lobbied to keep timber from federal lands off the market (to keep prices up), by the 1960s, the Forest Service was encouraging logging in national forests to meet the unprecedented demand. In essence, it fathered a major new sector of the industry, one born with a dependency on publicly owned old-growth. The federal government became a major supplier to the timber industry.

By the early 1980s, virtually all the lowland old-growth (except for the small amount within national parks and wilderness areas) was either gone or scheduled for logging. Then a major economic recession cut housing starts, timber demand, and prices. The industry responded by laying off many thousands of workers and requesting release from contracts requiring it to pay for trees it had once lobbied to cut. Yet even at a time of low demand, the cut on federal lands was greater than could be sustained: In national forests between 1980 and 1985, for example, the cut was 61 percent greater than actual growth of sawtimber.

As the economy picked up and housing starts increased, demand for timber and prices increased. The industry lobbied Congress to increase timber sales on federal lands. And, again, it got what it wanted.

III
The Keys to Understanding

Succession: Forest Birth and Rebirth

Phoenix . . . Egyptian mythology: a bird that consumed itself by fire after 500 years and rose renewed from its ashes.
(*The American Heritage Dictionary of the English Language* [1976])

On a breezy August day during a dry year, a rare Cascades lightning storm ignites a fire that sweeps through the giant conifers, lightly charring some, thrusting tongues of flame into the crowns of others. Mice, wolves, and woodpeckers flee, but trees, land snails, and salamanders cannot. For many, the fire ends their lives. But it is also a beginning, the setting for a new phase of life.

In a year, most burned spots are strewn with plants. Some sprouted from underground parts deep enough to avoid baking. Others germinated from buried seeds. Still others came from seeds borne by the winds or deposited in the droppings of sparrows, bears, and foxes. And the pioneering plants are not alone. Caterpillars chew their favorite leaves, oblivious to predatory wolf spiders coursing through the burn until it is too late. Creeping voles scurry among scorched fallen logs. Out of sight, pallid fungal threads colonize the sudden wealth of dead roots, bark, and wood.

In ten years, slim red alder and Douglas-fir saplings race to establish claims on the sky and the earth. Flocks of juncos and herds of black-tailed deer grow fat on abundant seeds and browse.

In fifty years, the winners are emerging, as the alders languish in the deepening shade of the Douglas-firs. Networks of fungal mycelia in the dead alder trunks send fruiting bodies through the bark to release powdery spores by the million.

In 120 years, the forest is dark and quiet. A dense Douglas-fir canopy intercepts so much light that only a handful of plants eke out an existence on the forest floor. Saplings suppressed by more vigorous competitors are unable to repel pathogenic fungi. Soon other fungi, the decomposers, will have their turn.

In 250 years, the largest Douglas-firs are producing substantial amounts of strong, fine-grained, rot-resistant heartwood. Many others, weakened by insects or diseases and toppled by storms, lie decaying on the forest floor. In shaded spots, the logs form nurseries for tiny, slow-growing hemlock seedlings. In sunny light gaps, however, the hemlock saplings on their nurse logs race quickly toward the sky.

In 500 years, huge, broken-topped Douglas-firs tower over giant western hemlocks, western redcedars, and smaller understory trees. The forest floor is thickly strewn with downed logs colonized by other plants. As the sun's last rays reach a den in a tall dead tree, a pair of flying squirrels awaken and descend, wary for spotted owls as they sniff the duff for truffles. Bands of Roosevelt elk wend their way among vine maples to drink from a cool stream.

In 1,200 years, the day after a storm toppled the last great Douglas-fir, hemlocks and redcedars now prevail. A snuffling black bear gorging on beetle grubs and salamanders rips through a rotting log, scattering panicking springtails and sowbugs. Ignoring other logs covering a fourth of the forest floor, the bear spares the next generation of hemlock seedlings growing slowly upon them.

The burn has undergone succession.

Succession is a change over time in an ecosystem's species composition, structure, and biogeochemical processes. It occurs when species establish and claim newly available habitats and are, in turn, succeeded by other species until directional change in species composition stops. Thereafter, species fluctuate but, in the absence of a disturbance, usually return to some average abundance. Individuals either replace others of the same species or alternate with members of other species, but the net effect is the same: a similar species composition from generation to generation. The ecosystem's physical structure and biogeochemical processes change as well.

Forest succession is not easy to study. In the frenetic world of bacteria, a researcher can chart succession in 100 petri dishes in a desktop incubator in a week. But in the stately world of giant conifers, studying successional processes is complicated by the fact that major species can live for more than a millennium, longer than twenty-five researchers' careers or 500 research grants.

As a result, piecing together successional sequences is more like archaeology than microbial ecology: We observe what is happening today and infer longer-term processes by comparing stands of known age in different stages. Under these circumstances, it should not be surprising that there are uncertainties about forest succession.

Because succession is integral to every aspect of forestry—including hydrology, recreation, timber, and wildlife and fisheries management— an understanding of succession is *the* most important basis for choosing among myriad, often conflicting options for managing forests. Every decision—to preserve or sell, thin or burn, spray or cut—hinges on succession. But people with different orientations have different notions about succession. Without a more comprehensive understanding, there can be no lasting agreement among loggers, conservationists, federal land managers, and congressmembers. This book is really a look at Westside coniferous forest succession, in which ancient forests are just the later, most threatened phases.

The first species to establish constitute a *pioneer community.* Pioneer communities and any later stages that are succeeded by still others are called *seral communities.* In contrast, a community that can indefinitely replace itself is a *climax community.* It can take 750, 1,000, or more years for a Westside forest to become a climax forest.

Succession begins when a resource becomes available. It can be a fallen leaf, the carcass of a deer, or soil exposed to the sun by a landslide. Because new resources are always becoming available somewhere or other, many species are adapted for locating and colonizing them. This allows individuals to reproduce before the resources are no longer available. Such species are opportunistic, specialists at taking advantage of opportunities.

Some opportunistic species, such as black cottonwood trees, have evolved exceptional mobility. Their myriad cottony seeds drift long distances on the winds. Most land in places where they cannot germinate. But the few that settle in suitable habitats—relatively bare, moist soil—can germinate quickly and race to pass on their genes. Other opportunists have evolved the ability to lie dormant as seeds, eggs, or spores until resources become available. Snowbrush seeds apparently persist in soil for centuries until a fire provides the heat and sunlight they need to germinate.

Despite their speed in locating new resources, opportunistic species are usually vulnerable to other species. Many are not very capable of fending off predators, parasites, diseases, or other species competing for resources. As their reproduction slows or stops, opportunists are replaced by species that can better withstand biological interactions. Although slower at locating new resources, these species are better at wresting control or holding onto resources as the community approaches climax. Some species, such as western hemlocks, serve as both colonizers or climax species under different circumstances. Random chance is important in determining which species colonize first.

The changing species composition and community structure alter physical conditions. As the forest canopy closes, less solar energy strikes the ground, where temperatures vary less, winds are weaker, and the soil surface retains more moisture. In general, physical factors become less extreme. These changes disadvantage opportunists, which are succeeded by species that do better under the new conditions.

Other changes are common during succession. Biomass (the amount of living and once-living material) tends to increase, the ecosystem's structure becomes more complex, and detritus (decomposing once-living material) becomes more important. The system becomes more efficient at cycling nutrients internally rather than allowing them to leak from the system. And, freed from the need to cope with physical extremes, species tend to form more intricate relationships with other species.

Succession involves not only major structure formers and photosynthesizers such as trees, but also the countless species that depend on them. Most animal, plant, and fungal species reproduce better in some successional stages than others. Indeed, some that depend on a particular stage are threatened because the stage has become rare. Oregon silverspot butterflies require early successional meadows and disappear when their habitats succeed to forest. At the other end of the sequence are northern spotted owls, which require old-growth.

Forest succession usually begins when some kind of disturbance has made resources available. It can be a fire, a landslide, a windstorm, an insect outbreak, or a timber operation. When a disturbance removes enough forest canopy, opportunistic species take advantage of the sunnier, breezier, drier, more open conditions and the more readily available soil nutrients and water. Plants needing high light levels to photosynthesize, fungi needing wind to disseminate their spores, and animals needing open spaces to watch for prey do better. Grasses, herbs, and shrub and tree seedlings are common colonizers. Red alders establish and grow quickly on moist and mesic sites, adding soil nitrogen as they do. Snowbrush has a similar role on drier sites.

Douglas-firs also colonize disturbed sites, and grasses, shrubs, and seedling broadleaf trees compete with them for water, nutrients, and light. Competitors sometimes exclude Douglas-firs for decades. Douglas-firs, however, grow far taller, and if they can reach the canopy, they outgrow, shade, and outlive their competitors.

Westside forest succession is typical in some ways but unusual in others. Conifers are usually climax dominants in the boreal taiga forests that form vast east-west bands across Canada and the USSR. In most temperate forests, conifers are seral; the climax trees are broadleaves, but in much of the Westside, broadleaves are colonizers, and climax dominants are conifers.

Further, seral trees are typically shorter-lived than the species that succeed them, which favors rapid succession and widespread climax communities. The Westside is different. With a maximum life span of 1,200 years, Douglas-firs, the region's most important seral species, live far longer than most climax trees that replace them. Still, like most early colonizers, Douglas-firs are not very tolerant of shade, and their seedlings do not grow well beneath intact forest canopy. In the absence of disturbance, Douglas-firs usually do not replace themselves in Westside forests.

At lower elevations throughout most of western Washington and Oregon, the most common tree seedlings occurring under old-growth Douglas-firs are western hemlocks. Full-grown hemlocks would be considered ancient giants most anywhere else in the United States, but they are smaller and shorter-lived (to 500 years) than Douglas-firs. Beneath the canopy, hemlocks grow slowly at best. Many die. Like seral species, hemlocks and most other climax species grow best at light levels brighter than deep shade. But unlike seral species, their seedlings and saplings can tolerate low light levels until the trees that shade them die. Then more light reaches them and their growth accelerates. The ability to "hang on" in shade allows them to succeed Douglas-firs sooner or later.

At higher elevations, where snowpacks are deeper, pioneering Douglas-firs or noble firs are often succeeded by Pacific silver firs, which reach similar heights and ages as western hemlocks. Western hemlocks and Pacific silver firs can form nearly pure climax old-growth stands, but some forest ecologists suspect that many "climax stands" never went through a Douglas-fir stage.

Western redcedars can also establish in shade and can live 1,200 years, perhaps much longer. But except in spots that collect standing water, they are usually interspersed with other trees.

In the warmer Siskiyou and Klamath mountains, where summer droughts are longer and fires are more frequent, successional sequences depend more on soil characteristics and exposure. Although Douglas-firs often predominate, the forests differ from those of the Olympics, Coast Ranges, and Cascades. Ecosystem diversity is higher, as is the diversity of tree species within ecosystems. Pacific silver firs are rare or absent; western hemlocks and western redcedars occur only on cooler, moister sites.

In this area, evergreen broadleaves such as tanoaks, golden chinquapins, and madrone are more tolerant of dry conditions than Douglas-firs, not less. Near wet spots, Port Orford cedars can be important; incense cedars and sugar pines are sometimes codominant with Douglas-firs in seral stands. The driest sites can even have ponderosa pines. Climax forests tend to be dominated by white firs or tanoaks.

Elliot A. Norse

Western redcedar seedling, Mt. Baker-Snoqualmie National Forest (Washington). *Redcedars, Alaska-cedars, western hemlocks, and some firs can survive in the shade beneath the forest canopy for many decades, but their growth is very slow. When a big tree dies (opening a gap in the canopy), more light reaches the forest floor and trees there grow much faster.*

Along the coasts of Washington and Oregon, giant, long-lived Sitka spruce dominate. Because there is so much rain and fog—these are truly temperate rainforests—red alders and bigleaf maples can establish on all but the driest sites. Douglas-firs are less common than in drier, more fire-prone sites inland. Sitka spruce can establish in both lighted and shady spots. Western redcedars are more abundant than in inland sites, and western hemlocks are everywhere.

Despite the variations in different subregions, some general patterns emerge. The first trees to establish grow upward and outward quickly, so the canopy rises and coalesces. As trees grow, older branches are shaded and carry on less photosynthesis, but their metabolic costs do not decrease. Old branches become a liability, die, are attacked by decomposer fungi, and eventually fall to the ground.

Whole trees are subject to the same laws. When seral trees are outgrown, their energy budgets run "in the red." Because defenses against fungi and insects are metabolically costly, these trees become vulnerable when they can no longer afford to defend themselves.

In the competitive world of a younger forest, canopy coverage approaches 100 percent, and the forest floor is littered with needles, twigs, branches, and logs, but little else; it is too dark. The small gaps in the canopy created when trees die are quickly filled by lateral growth from their neighbors. The number of trees per unit area diminishes, but the crowns of the trees get wider.

To someone familiar with forests elsewhere, the crown geometry of old Westside conifers is distinctive. The trees seem very tall and skinny because they have limited ability to grow laterally. Hence, after they reach their maximum crown diameter, the death of a neighbor creates a gap that remains for decades until it can be filled from below. In these gaps, more sunlight reaches the forest floor. In old-growth stands, light gaps are larger and longer-lived than in young stands.

The uneven light levels shape the forest's structure, species composition, and functioning. Well-developed herb and shrub layers appear in sunny patches. Shade-tolerant understory trees of various heights fill space below the canopy. These plants provide a far more complex vertical structure than occurs in a young or mature forest, and the patchy distribution of gaps creates more horizontal diversity. Indeed, gaps and trees of different sizes give an ancient forest canopy a characteristic ragged appearance from the air.

In shady spots, growth is slow even for shade-tolerant species. But as stands age and gaps last longer, the energy income of shade-tolerant trees improves, allowing them to grow faster and reproduce. Eventually, they become the canopy trees, and their seedlings and saplings await their turns in the still, shaded world below.

While the structure of living vegetation is becoming more complex, large trees that die are also creating large snags and fallen logs. The lighted patches, shrubs, slow-growing species, large snags, and logs provide habitat for species that were scarce or absent in the mature forest. Time and tree death have created an old-growth forest.

If all disturbance stopped, plants that had colonized disturbed sites would be replaced by later seral species until climax communities prevailed everywhere. Succession would cease.

There is no great risk of this happening. Disturbances happen on all spatial and temporal scales. They range from volcanic explosions more violent than the largest nuclear blast to the footfalls of deer. Some are common, others rare. Some happen randomly, others periodically. All shape the successional pattern of a landscape.

Many kinds of disturbances—floods, volcanic eruptions, mass slumping, avalanches, ice storms, winds, lightning strikes, fires, diseases, impacts from other falling trees, attacks by defoliating or boring insects, and logging—kill trees in the Northwest. The different kinds have differing effects. For example, both avalanches and volcanic pyroclastic flows can snap trees like toothpicks and bury them. But burial under snow has rather different implications for recolonization than burial under glowing-hot volcanic debris.

Some disturbances reset the successional sequence by killing virtually everything. Others are selective, setting succession back only partway because most things survive them. For example, insect pests and diseases (and some foresters) select trees of certain species and age classes but not others. Fire can exert selection as well. Older trees of some species, such as Douglas-firs, have thick bark that resists fires better than younger individuals and species with thin bark, such as firs, western hemlocks, and western redcedars. Some disturbances, such as ice storms and fires, are seasonal, favoring colonizing species whose reproduction coincides with the disturbances.

Sites have differing vulnerabilities to floods, mass slumping, lightning strikes, and fires. And some kinds of disturbance occur over areas large enough to kill trees by the million, while others usually kill individuals here and there.

Indeed, the speed with which an ecosystem recovers depends on the size of the disturbance. Small disturbances (such as individual treefalls) set the successional sequence back less because small disturbed patches have microclimates similar to surrounding undisturbed areas and because equilibrial species that might recolonize are nearby. In larger disturbed areas, however, conditions are more severe and colonizers are farther away, circumstances favoring opportunistic species, which are better at long-distance dispersal.

Because disturbances vary so much in space and time, natural landscapes are not uniform climax communities. Rather, they more closely resemble quilts or mosaics whose irregular-shaped patches are communities with differing structure, species composition, and ecological functioning.

Landscapes are heterogeneous for other reasons as well. Because weather varies from year to year, precisely *when* a disturbance occurs can be important. For example, foresters are often frustrated in their attempts to regenerate forests like the ones they cut, especially in drier areas. But if the original forest was established when a disturbance was followed by unusually wet years, a comparable stand might be impossible to establish until equally favorable conditions prevail. Mosaic landscapes are created by both average conditions and by uncommon events that favor some species at the expense of others.

Dead Trees: The Life of the Forest

OLD TREES AND YOUNG TREES

Shibui *(Japanese): beauty of aging (adj.). . . . Think of Katharine Hepburn's face. She was beautiful as a very young woman. And she grew into a different kind of beauty as age revealed the grain of her character that was hidden beneath the smoothness of her youthful face. . . . Perhaps, as those of the baby-boom generation move into their forties and fifties . . . a renewed appreciation of this kind of beauty will bring new attitudes.*
(Howard Rheingold, *They Have a Word for It* [1988])

Most animals have short life spans, a year or less. Humans, unusual in so many ways, are in the company of a few species—Asiatic elephants, Andean condors, box turtles, lake sturgeon, and brain corals—as the longest-lived animals on Earth. Still, our life spans of seventy or eighty years, 120 at most, seem all too short. For many peoples and religions, why death comes so soon is a central mystery of life.

Although many plants are also short-lived, some survive centuries or more. The western United States is particularly rich in long-lived trees. At least one bristlecone pine has clung to a bleak, windswept peak in California's White Mountains for 4,600 years. California's giant sequoias and coast redwoods are longevous, as are giant Westside conifers. The oldest Alaska-cedar might have burst from its seed while Moses was leading the Israelites out of Egypt.

To be enveloped in silence and surrounded by columns of ancient life evokes in many people an image of great European cathedrals. But to many loggers and even some foresters, the trees of cathedral forests are merely "overmature," "senescent," or "decadent." Individualistic in form, bearing the furrows and scars of time, these oldsters seem wasteful, for they occupy ground that could be producing more wood to benefit humankind. They are not paying their way.

Young trees are different. Compared with the forms of the old, young trees are straight and symmetric, often so similar that they seem made with a cookie cutter. In a few decades—usually three to five in a Douglas-fir, more in slower-growing species—the amount of leaf tissue stabilizes. Still, the tree continues to grow, reaching maximum height during its second or third century. Like many of us, old-growth conifers continue to add girth long after height growth has ceased.

Giant Westside conifers are not the world's fastest-growing trees, but they grow for an exceptionally long time. Growth slows sometime during their second century as the energy cost of maintaining their increasing biomass rises while photosynthesis does not. The wood they produce is more valuable than that of younger trees—knot-free, fine-grained, strong, and decay-resistant—but the quantity is not as great.

As with humans, time takes a toll on the survivors. After enduring centuries of storms, fires, and insect attacks, many have lost their tops; their trunks are scarred, and they have rotten spots in the limbs, heart, or butt. They have become old-growth trees. But barring holocaust, gale, or timber sale, many can survive centuries longer.

All things being equal, young and mature trees add more wood per acre than old-growth trees; if the only concern is the quantity of wood produced, then old-growth is indeed wasteful. Therefore, it makes one kind of economic sense to cut ancient trees (especially since their wood is worth more) and to replace them with fast-growing young trees. But all things are not equal. In addition to producing more valuable timber,

Elliott A. Norse

Young and old Douglas-firs, Willamette National Forest (Oregon). *Young saplings and old trees contrast sharply. The symmetrical shapes of young and mature trees are so alike that they appear to have been made with a cookie cutter. Old trees, having endured centuries of wind, fire, disease, and insect attack, are individualistic, no two alike. The Japanese call this beauty of aging* shibui; *there is no comparable term in English. Many ecological differences between young or mature forests and ancient forests result from the different forms of the trees.*

old-growth forests provide many essential benefits that young and mature forests do not, not only in the centuries while trees are old but for centuries after they die. Indeed, this leads to a conclusion that might seem counterintuitive, but one that researchers find is true in many ways: *Dead trees are the life of the forest.*

This idea is anathema to some people. A common notion in our youth-oriented society is that forestry is about growing and cutting trees and that everything good about forests comes from their young, healthy, vigorous growth. But this misses a crucial part of the cycle. A Douglas-fir can live more than 1,000 years and as a downed log can influence the forest for hundreds more, yet some foresters focus solely on the period of highest growth that lasts only a century or two. Emphasis on this portion of the forest life cycle misses the contribution of the much longer period during which many other values reach maximums and the old-growth ecosystem prepares for the next phase of vigorous productivity.

Simplistic emphasis on growth is apparent in timber industry ads extolling the virtues of planting millions of vigorous, genetically "improved" seedlings. It is evident in wildlife managers' efforts to produce ever greater biomass of a selected few game species at the expense of other species. And it is unmistakable in the pejorative application of the term "dead wood" to people who are considered unproductive and useless. Yet there are strong economic, environmental, and scientific reasons for maintaining our old-growth.

Although growing trees is important, inordinate emphasis on growth lies at the core of the controversy over forest management in the Northwest. Foresters have been taught that maximizing fiber production is the primary goal of forestry. But new scientific research is demonstrating how this fails to encompass the intricate complexity of forest ecology. Ecologists are discovering how each phase of succession contributes to the productivity of trees as well as to wildlife, climate, hydrology, and the shape of the land itself. We are learning how the seemingly contradictory phenomena of death and life are inextricably intertwined. Nowhere is this relationship more apparent than in the ancient forests of the Pacific Northwest.

TREE DEATH: THE BIRTH OF SNAGS AND LOGS

... there are some people who really enjoy old things. The unchanging, the stagnant, the dying. ... The longer a city stagnates or the longer a ruin stands, the more beautiful it becomes. It's kind of ridiculous, isn't it?

The Parthenon in Athens, for instance, was not even recognized as an object of beauty until 1,800 years after it was constructed. ...

Old growth forests are ... built on functions of death ... ecosystems that contribute little to others and are extremely self-centered in that they care for themselves and no one else.

(Robert Vincent, *Old Growth Forests, A Balanced Perspective* [1982])

A tree's first year is often its last, for a seedling's life is perilous. It must compete with other plants for light. Its root system cannot reach deep enough when drought robs moisture from the upper soil layers or when erosion carries it away. And a browsing deer can pluck it, leaving nary a trace.

Later years are less precarious, but, as Steve Cline and coauthors showed (1980), the risk of death is still high, especially for saplings stressed by competition. Unable to store enough energy to combat defoliating insects and pathogenic fungi, many succumb, leaving abundant slender snags and logs in dense young stands. The annual death rate declines further as surviving trees attain maturity and old-growth. Great old trees may hold their places in the canopy for centuries, but, eventually, they too will die. And while the death of a tree is a loss in one sense—it is no longer producing wood—it is also a vital beginning for the forest ecosystem.

It is not only *when* a tree dies, but *how* it dies that determines its role in the old-growth ecosystem, because some agents of death leave trees upright but others send them crashing to the forest floor. Bark beetle infestations, diseases, and some fires leave trees standing as snags until some later disturbance brings them down. Windstorms, landslides, and volcanic pyroclastic flows topple trees immediately.

Trees die for different reasons in different places. Jerry Franklin, Hank Shugart, and Mark Harmon (1987) noted that winds account for 83 percent of tree deaths in an Oregon coastal Sitka spruce/western hemlock forest, but Harmon and coauthors (1986) found that winds take only 17–47 percent in Cascades Douglas-fir/western hemlock forests. The actual cause of death is often complex. Trees weakened by adverse conditions are more vulnerable to defoliating insects. Trees weakened by defoliators are more vulnerable to invasion by some boring insects. Boring insects, in turn, provide entryways for disease-causing fungi. And trees weakened by fungi are more vulnerable to windstorms.

Tree mortality also fluctuates from year to year. For example, the number falling to the forest floor increased up to sixty times during a bark beetle epidemic in a 180-year-old Douglas-fir stand in Oregon. Fires might not touch a stand for fifty years or 500, but hot fires can create huge pulses of snags and logs. In fact, the fires that initiate most young Douglas-fir stands can produce more snags and logs than ancient forests do. But whatever the cause, whether the "life" of a dead tree begins as a snag or a downed log determines its role in the forest ecosystem.

Snags and logs are vital to ancient forests because they increase structural complexity, producing places for animals to feed, display, nest, take refuge from harsh weather, and escape from other species. A major reason is that they can be penetrated by some organisms but not others (such as their predators, parasites, or competitors). Penetrable substrates provide more kinds of shelters than impenetrable substrates. That is why coral reefs, which consist of penetrable limestone riddled with holes and cavities, have higher species diversity than reefs of impenetrable rock. In terrestrial ecosystems, the main penetrable substrates are soil and large pieces of decaying wood.

The sound wood of young, living trees is hard. As a result, forests provide homes for more species when they have large, old, rotten trees, snags, and downed logs, which claws, beaks, plant roots, and fungal hyphae can penetrate. Rotting snags and logs provide the tunnels, dens, and nesting cavities needed by animals from black bears and spotted owls to land snails and springtails. They are the birthplaces for western hemlocks, Sitka spruce, and smaller plants. They are habitats for many of the Pacific Northwest's remarkably diverse mushrooms. They are sites of biological nitrogen fixation, adding to the nutrient wealth of the forest. And they provide shady, moist refuges that speed recolonization of burns and clearcuts.

Dead wood? Yes. Useless? Hardly.

SNAGS AND DEAD TOPS

. . . the dead tree gives no shelter.
(T. S. Eliot, *The Waste Land* [1922])

Standing dead trees and fallen ones have different ecological roles. Snags offer sloughing slabs of bark, chimneylike cavities and vantages out of reach of ground-dwelling predators. They also attract birds whose droppings contain seeds of blackberries, salmonberries, thimbleberries, huckleberries, and other plants that colonize clearcuts and burns. Some plants, fungi, and invertebrates use both snags and logs, but most vertebrates use mainly one or the other.

Snags are particularly important for birds, bats, and carnivorous mammals. Raptors (hawks and owls) use them for lookout and plucking posts; many birds use them for courtship display, nesting, feeding, and roosting sites and for thermal protection at night, during heavy rains, and in cold weather. Loose sheets of bark on snags are major breeding sites for brown creepers and for several kinds of bats. Cavities at the bases of snags serve as dens for mammals as large as black bears; cavities higher up are nesting sites for 30–45 percent of Northwest forest bird species and dens for squirrels, martens, and bobcats.

Snag, Olympic National Park, and dead top, Olympic National Forest (Washington). *Standing dead trees are vitally important forest structures that allow many species to escape from predators and harsh weather. After insects and fungi soften the wood, woodpeckers excavate nesting cavities in snags. When woodpeckers abandon them, other birds and mammals move in. Cavities in snags are nesting sites for 30 to 45 percent of Westside forest birds. Dead tops are also important nesting sites for birds such as ospreys and eagles.*

In general, smaller snags can be used by some smaller species, but large snags are essential for larger species and some small ones too. Wood ducks, great horned owls, pileated woodpeckers, and fishers generally need snags more than 25 inches in diameter, but so do some smaller species, such as Vaux's swift. Height is also important; far more cavity-nesting species use tall snags than low-cut stumps. Although young forests often have more snags, the larger snags in ancient forests are far more valuable to cavity-dwelling wildlife.

Most cavities in hard (younger) snags are excavated by woodpeckers preparing to nest. Chickadees and nuthatches can also excavate cavities in soft (older) snags. When excavators move out, a wide variety of other birds and mammals can move in. By providing a crucial limiting resource, excavators function as keystone species, ones whose activities disproportionately influence other forest dwellers.

Most excavators, in turn, depend on insects and fungi to soften snags. Wood-boring beetles tunnel into trees soon after they die, providing entry for decomposing fungi. This allows woodpeckers and bears to shred the softened wood as they forage for beetle larvae. People who automatically consider insects and fungi to be pests might be surprised to learn of their essential service to all cavity dwellers, from furbearing mammals and gamebirds to rare species such as spotted owls.

As decomposition and shredding weaken a snag, it first loses its needles and twigs, then upper branches, bark, and the top of its trunk, then wood farther down the bole. Some snags are firm when they fall, but many just disintegrate and collapse. Size affects their longevity; small ones decompose faster. In western Oregon, small Douglas-fir snags characteristic of eighty-year-old stands fall within two decades. The large Douglas-fir snags of ancient forests usually last more than 125 years. The species of snag affects longevity as well. Western redcedar and Douglas-fir snags outlast western hemlocks and silver firs. So, diameter, height, species, degree of decomposition, and numbers all affect the value of snags to forest wildlife.

Entire trees need not die to provide essential vertical dead wood habitat for forest wildlife. The tops of many ancient conifers are dead, victims of winds, freezing, lightning strikes, bark beetles, or pathogenic fungi. Although dead tops—often the top 10 percent, sometimes even the top half—are difficult to see from the ground (and, hence, are often missed in surveys), canopy dwellers use them the same ways that they use snags.

DOWNED LOGS

It may take centuries for a large nurse-log to completely disintegrate, by which time the trees it fostered will become giants themselves, standing hollow-rooted in a colonnade. There are few places in nature where the eternal cycle of life, death, and rebirth is more strikingly apparent.
(Pat O'Hara and Tim McNulty, *Olympic National Park: Where the Mountains Meet the Sea* [1984])

The expressions "like a bump on a log" and "sleeping like a log" reflect the common view that logs are not very lively. But like many widely held ideas, this one misses the mark. Large living trees actually consist mainly of nonliving wood and bark, whereas downed logs are hotbeds of biological activity.

And logs can certainly be abundant. On first seeing an ancient forest,

Tim Crosby

Downed logs, Mt. Baker-Snoqualmie National Forest (Washington). *Like snags, these dead trees are hotbeds of biological activity. They are highways for small mammals, homes for salamanders, reservoirs for mycorrhizal fungi, and birthplaces for trees. They are vitally important lifeboats during wildfires and logging. Westside ancient forests have more large downed logs than any other forest ecosystem.*

after the pervasive greenness and the towering trees, the visitor next notices logs strewn about like titanic pickup sticks. They cover up to 25 percent of the forest floor in some old-growth Douglas-fir/western hemlock stands, more than in any temperate forest ecosystem. And they constitute a substantial fraction of the wood even in an ecosystem renowned for extraordinary biomass. One 515-year-old stand had 490 metric tons/hectare (219 tons/a.) of large logs. With snags adding up to another 105 tons/hectare, dead trees are clearly major biomass components in ancient forests.

Not surprisingly, living things make use of this prodigious resource, as beautifully depicted in the Forest Service's *The Seen and Unseen World of the Fallen Tree* (by Chris Maser and coauthors [1984]). Fallen logs provide shelters for shrews and mice. Logs that have fallen across-slope are especially valuable because they accumulate rich, loose soil on the uphill side, in which small mammals can burrow, and overhangs on the downhill side, which serve as refuges.

A new log is colonized by insects, particularly beetles. They bore through the outer bark to get at the inner bark and sapwood, tissues with higher nutrient concentrations and lower amounts of defensive chemicals. Because outer bark and heartwood are less attractive to decomposers, they decompose more slowly.

By initiating decomposition, thereby providing ingress to the many species that use fallen logs, beetles function as keystone species. Most (if not all) are mutualistic with fungi and carry spores with them. The fungi decompose the inner bark, and the tunneling beetles often eat the fungi. As tunnels spread, carpenter ants, termites, millipedes, mites, spiders, salamanders, and the roots of plants gain entry.

Logs decay so predictably that ecologists have developed a system for classifying them. Class 1 logs are mostly intact and undecayed. They become Class 2 logs when their small twigs fall off and inner bark and sapwood are invaded by decomposers. In Class 3, plants are growing on the upper surface, roots of others have invaded, the outer bark has sloughed off, and the log sags where it crosses obstacles. By the time a log reaches Class 4, heartwood has begun to collapse into an oval mound of soft reddish blocks. It takes a good eye to spot Class 5 logs, which have degraded into finely divided crumbly organic material, merged with the forest floor, and been covered with plants. Decomposers quickly convert the resource-rich Class 1 logs to Class 2; succeeding stages last progressively longer.

Downed logs retain more moisture than snags, in both their decaying wood and the soil beneath, and serve as bridges and runways for small animals on the forest floor. As with snags, large logs last longer than small ones, because they have more decay-resistant heartwood and dry more slowly; wet wood excludes oxygen and slows decomposition. Similarly, redcedar and Douglas-fir logs last longer than hemlock and fir logs. But logs last longer than snags. Less exposed to drying winds and having more contact with the soil, logs stay wet and decompose more slowly and can last two centuries or more.

The nutrient content of a log increases with time. Tree boles—trunks and bark—are poorer in most nutrient elements than leaves, twigs, and other plant parts. During decomposition, however, as carbon compounds oxidize and disperse as carbon dioxide, concentrations of other elements increase. And they increase for two other reasons. Nutrient-rich materials from above fall onto logs and decompose, and bacteria within logs enrich their surroundings by fixing nitrogen.

Logs are crucial for reproduction of some trees. In coastal Oregon, Mark Harmon (1987) of Oregon State University found that very few western hemlock and Sitka spruce seedlings can establish on forest floors covered with thick mats of competing mosses. Rather, nearly all recruitment occurs on logs, but not just any logs.

It might seem that older logs are best because decomposition releases their nutrients. But the opposite is true; most hemlocks and Sitka spruces get their start on newer logs because later stages get covered by the same mosses that prevent seedling establishment on the forest floor. Naked bark is not very hospitable unless it is covered with decomposing twigs and leaves. But when it is, seedlings have moist humus from which to extract nutrients until their roots reach mineral soil. Then, with more abundant nutrients and more constant soil moisture, seedlings can grow as fast as available sunlight allows.

Downed logs harbor diverse fungi that form mutualistic associations with plant roots called mycorrhizae. Mycorrhizae enhance the conifer's uptake of water and nutrients and protect it from soil-borne pathogens and (perhaps) competing plants.

When catastrophic disturbance visits a stand, logs can be important to reforestation, as revealed by the research of Jim Trappe, Chris Maser, and co-workers. In many cases, disturbed forest soils support fewer mycorrhizal fungi, slowing conifer reestablishment. But older downed logs are like sponges. Even during summer drought, they are often too wet inside to burn. They serve as a vital legacy of the former forest by sheltering organisms, such as mycorrhizal fungi, that speed forest re-

Elliott A. Norse

Western hemlock and Sitka spruce seedlings on decomposing log, Olympic National Park (Washington). *Most western hemlock (left, bottom, far right) and many Sitka spruce trees (center) grew from seeds that germinated on decomposing nurse logs, one reason for the saying "Dead trees are the life of the forest."*

covery. Seedlings that germinate nearby have a ready fungal inoculum, hence improved growth and survival. And there is a tantalizing suggestion that big logs disseminate mycorrhizal fungi farther afield as well by serving as lifeboats for small mammals.

A newly logged or burned site is an inhospitable sea for western red-backed voles, deer mice, Townsend chipmunks, and other small mammals. Without hiding places, they are easy prey for snakes, hawks, and foxes. But beneath fallen logs, they find shelter and food as well: hypogeous (underground-fruiting) fungi, many of them mycorrhizal. When the small mammals leave their refuges, they disperse viable spores in their fecal pellets, seeding the disturbed areas with mycorrhizal fungi.

Logs are even more important in the streams that drain ancient forests. From small headwaters through medium streams, logs shape the streamcourses by forming dams and baffles that modify flow patterns. In the absence of logs, medium streams tend to consist mainly of fast-flowing runs having bottoms of well-graded sediments, providing good habitat for a few species (such as steelhead trout less than one year old) but not for many others (such as chinook salmon parr).

In contrast, streams crisscrossed by logs have quiet pools interspersed with shallow riffles. Currents rushing against logs slow abruptly and lose the ability to carry coarse sediments and, therefore, accumulate them in pockets. Thus, logs determine the pattern of sediments on the streambed, from cobbles down to silts.

Logs increase stream habitat diversity, providing opportunities for more kinds of living things. Olympic salamanders that would be swept away by currents can find quieter riffles in which to forage. Salmon parr can choose places where they can minimize energy expenditures from swimming while maximizing their energy income in food insects that currents bring. Bottom-dwelling insects can select among substrates from fine sediments to submerged logs.

Universal truths—whether in human relations, economics, or ecology—span both the best and the worst of times. So it is with the value of logs in ancient forest streams. When big winter storm systems cover the Westside with rainy gray curtains, streamflow rises precipitously. Gurgling streams become cascading torrents strong enough to lift boulders and rip living things from their homes, swirling them downstream to their deaths. Small logs are often swept downstream with them. But logs large enough to resist the currents provide crucial refuges of slower water, in which stream dwellers can avoid the scouring flow. More salmonids survive storms in ancient forest streams than in streams that flow through areas deficient in logs, such as tree plantations.

The death of something valuable, such as a giant conifer, seems wasteful to many people. It is difficult to let a tree that could yield thousands of board feet become nothing more than a refuge for voles and truffles, salamanders and salmon. But scientists are learning that logs are the "hot-spots" of ancient forest ecosystems, essential to biological diversity and sustainable productivity. Such trees are dead, true enough, but theirs is a lively death indeed.

LITTERFALL AND THE DEATH OF ROOTS

Nature does nothing uselessly.
(Aristotle, *Politics, Book I* [4th century B.C.E.])

On a walk through an ancient forest, it is hard not to notice yard-wide snags towering 175 feet and fallen logs of equal stature. Dead trees provide much of old-growth's unique vitality. But death also propels the pulse of forest life in two other ways. One of them is easy to see if you know what to look for. The other is hidden in the grainy darkness of the soil.

Find a moss-covered log, sit quietly, look around, and it should not be long before you see something fall from the trees. It can be a green branchlet snipped by a foraging chickadee or a dead one snapped by an alighting chickadee. It can be a flake of bark sheared by a downy woodpecker, a lichen dislodged by a gust, a ghostly insect exoskeleton, a wrinkled caterpillar frass pellet, or even a feather from a spotted owl.

Some falling material is so fine that it will be missed unless the light is exactly right. But visibly or not, the old-growth giants are showering you with their sifting pollen, slowly settling fungal spores, and sinusoidally descending strands of spider silk. At other times, gales snap treetops and large branches, sending them crashing.

This varied rain of material from the trees bears the unpoetic name "litterfall." In a sense, litterfall is the capillary system of the forest, removing wastes and conveying food to the legions of consumers down below. Although far less conspicuous than the giant logs, the fall of litter is a primary path by which nutrients are cycled among the living things in a forest ecosystem.

Aside from the carbon, oxygen, and hydrogen that make up 95 percent of plant tissue, some thirteen nutrient elements are essential to plant growth. You can get some idea of the role of litterfall in forest nutrient budgets by following what happens to one crucial element, nitrogen.

The nutrients available to a plant come from three places. In the short term, most are recycled, coming from decomposing biomass. But there are two ultimate sources: rocks and air. Nitrogen is unlike other nutrients in that virtually none comes from weathering rocks.

The air contains chemicals that settle in ancient forests as microscopic particles during dry weather or in fog, rain, and snow. Their origins are diverse. Capricious winds wrest sodium and chloride ions from foam-tipped waves and loft them into the atmosphere. Smoke from burning forests in Madagascar and Borneo wafts magnesium and boron to the slopes of Mt. Baker. Sulfur belched from China's coal-fired power plants restocks the chemical larders of ancient forests. Volcanoes from the Cascades to Kamchatka deposit stygian nutrients in dustings of ash.

Of course, nutrient deposition from air varies greatly. In England, up to 14 kg of nitrogen per hectare (12.4 lb./a.) sifts from the air each year. Pacific Northwest air is much cleaner, but still deposits about 1–2 kg/ha.

The canopy of ancient forests is ready to capture this fallout. The canopy is a remarkable filtering system. As University of Oregon researcher Lawrence Pike and his coauthors (1977) discovered, a single 400-year-old, 77-meter (253 ft.) tall Douglas-fir can have 60 million needles; the surface area of canopy needles can be fifteen times that of the ground surface. New needles are fairly clean, but older ones are covered by a scuzzy film of microscopic fungi that absorbs and transforms chemicals borne by the atmosphere. And, as in tropical rainforests, twigs, branches, and trunks of old-growth trees and snags are festooned with communities of epiphytes (nonparasitic plants and fungi that grow on trees). Epiphytic green algae, mosses, liverworts, fungi, lichens (in the wettest Westside forests, the Sitka spruce/western hemlock rainforests), clubmosses, and ferns obtain nutrients from the air and decomposing litterfall. Some of the epiphytes—notably lichens in the genus *Lobaria*—add to the ecosystem's nitrogen pool in a second way, by chemically "fixing" atmospheric nitrogen gas. Together— growing, competing, reproducing, and dying—these "air plants" amass substantial amounts of soil on large branches far above the ground.

When it rains, some of the canopy's nutrients dissolve and drip through the foliage (throughfall) or form thin soupy rivulets down the trunks (stemflow). Epiphytes intercept some, but throughfall alone provides perhaps 3 kg/ha of nitrogen to the soil each year in an ancient Douglas-fir stand.

Another way that nutrients in the canopy return to the forest floor is through grazing by creatures from tiny mites to lichen-munching northern flying squirrels and red tree voles that feast on Douglas-fir needles. These fungivores and herbivores are eaten by predators from spiders and warblers to the canopy's top carnivores: pine martens and northern spotted owls. In pellets and poops, grazers and predators send nutrients to the forest floor. A much larger fraction of canopy foliage and epiphytes falls uneaten.

Whether or not it has passed through an animal, litterfall contributes large amounts of nutrients to the soil. One Oregon old-growth Douglas-fir forest sends a yearly average of 32.7 kg/ha of nitrogen to the forest floor. Litter from other species can be even richer; nitrogen-fixing red alders in British Columbia add about four times as much. Even within a stand, various kinds of litter have differing nutrient contents. In an Oregon old-growth Douglas-fir stand having 11 metric tons/hectare of annual litterfall, only 26 percent is foliage, but it contains 68 percent of the nitrogen.

When litter reaches the forest floor, it is attacked by decay fungi, and bacteria. They take about a decade to release all the nutrients from fine litter; coarser litter takes longer. The nitrogen in a year's litterfall might seem inconsequential compared with a 100–200-year supply in the soil. But over centuries, the steady organic rain becomes significant in the nitrogen budget of an ancient forest.

In the last decade, ecologists have found that soil nutrient dynamics can be even more dynamic than we had suspected. In snowy montane Pacific silver fir forests, litter decomposes very slowly, so a significant fraction of the ecosystem's nutrient capital is "tied up." Trees respond by producing large amounts of fine roots in the duff and upper soil layers. Underground production is about *four times* as great as aboveground production. Even in warmer Douglas-fir forests, where decomposers release nutrients faster, production of fine roots and mycorrhizae can equal aboveground production.

Ecologists are still in the early stages of learning about the basic processes that sustain forest productivity because some very important ones occur where we had not looked, out of sight and out of reach: above our heads and below our feet.

The Definition of Old-Growth

What's in a name? That which we call a rose by any other name would smell as sweet.
(William Shakespeare, *Romeo and Juliet* [1595])

How something is defined can shape the whole debate on an issue. Indeed, a definition can serve as a shibboleth. The ways that people define old-growth reveals what they consider important. Those that see forests as fiber factories define them in terms of wood production.

Seedlings in a new burn or clearcut produce no marketable wood at first but become commercially usable in a few decades. The wood they add per acre increases annually until reaching a peak. Timber-oriented foresters consider this "culmination of mean annual increment" the transition between young (too young to cut) and mature (ready to cut) forest. Westside Douglas-fir stands generally reach commercial maturity at 80–110 years, although companies cut them at 50 or even 40 years when timber prices are high.

Because determining when a stand reaches maximum annual wood production takes time and effort, foresters often define the maturity of Douglas-fir in a way that is far easier to measure: when trees are at least 21 inches in diameter. Depending on site quality, trees can reach this diameter before or after culmination of mean annual increment. But these definitions have nothing to do with biological maturity, that is, age of first reproduction. Westside conifers begin producing seeds decades before they are commercially mature.

Further, mature and young Douglas-fir stands are ecologically similar in some important ways. They both have fairly evenly distributed trees that germinated after disturbance eliminated the previous stand. They both have sparse or absent undergrowth, closed canopies, trees with conical tops, and a cylindrical form. To an ecologist, commercially mature forests differ from young forests mainly by having fewer snags and logs, by having fewer and bigger trees, and by being less vulnerable to fire.

Mature stands continue to add usable wood at a gradually declining rate. For patient foresters, this decline is offset by the increase in the quality of the wood they add—fewer knots, finer grain, greater strength, and decay-resistance. But as trees age, many lose their tops in storms and develop rotten spots in their trunks. Foresters term them senescent, decadent, overmature, or old-growth.

As discussed in the earlier section on succession, stands undergo many other ecological changes as they age. They begin to acquire these old-growth characteristics as early as 175–200 years, although old-growth is still "young" at 250 years and changes continue to accumulate. As Tom Spies and Jerry Franklin (1988) point out, some old-growth characteristics do not peak until a stand is at least 500 years old. A key process is the death of large trees, which creates the snags, logs, and long-awaited dapples of sunlit opportunity for understory trees and shrubs. The diverse trees, shrubs, large snags, and logs provide habitat for species that were scarce in the mature forest. Concurrently, slow-growing species have an opportunity to mature and reproduce. Time and tree death turn a closed-canopy forest into an ancient forest.

So, wood production and ecological characteristics provide alternative ways to view old-growth. One is timber-oriented; the other is forest-oriented. How does the Forest Service define old-growth? Table 3.1 gives definitions in the eleven draft Westside national forest management plans issued to date (all but the Klamath in California).

Note that only Mt. Baker-Snoqualmie, Olympic, and Willamette national forests define old-growth the same way and that their definition is

Table 3.1

OLD-GROWTH FOREST DEFINITIONS IN WESTSIDE NATIONAL
FOREST PLANS

National Forest	Definition
Mt. Baker-Snoqualmie (Wash.)	Mature large sawtimber with diameter breast height (dbh) 21 in. or greater
Olympic (Wash.)	Stands with dbh 21 in. or greater
Gifford Pinchot (Wash.)	Stands 250 years of age or older
Mt. Hood (Oreg.)	Stands 200 years of age or older
Willamette (Oreg.)	Stands with dbh 21 in. or greater
Umpqua (Oreg.)	Acres of forestland which have no record of harvest activity
Rogue River (Oreg.)	Stands 10 acres or more in size with mature and overmature trees, a multilayered canopy and several age classes, and no significant evidence of human activity
Siskiyou (Oreg.)	Stands 10 acres or more in size showing mature and overmature trees, multilayered canopies and trees of several age classes, dead and downed woody material, and no significant human impact
Siuslaw (Oreg.)	Stands over 10 acres containing four or more large old trees and a younger understory
Shasta-Trinity (Calif.)	Old-growth not separated from older mature stands with tree height exceeding 50 ft. and canopy cover greater than 40 percent
Six Rivers (Calif.)	Stands of mature and overmature trees with dbh 21 in. or greater and tree height exceeding 50 ft.

SOURCE: Anderson and Kloepfer (1988).

timber-oriented. Others emphasize age (e.g., Gifford Pinchot), forest structural characteristics (e.g., Rogue River), history (e.g., Umpqua), or combinations of these (Siskiyou). Six of them are based on a single criterion. In seven (Mt. Baker-Snoqualmie, Olympic, Willamette, Rogue River, Siskiyou, Shasta-Trinity, and Six Rivers), mature and old-growth forests are combined. Only three plans (Rogue River, Siskiyou, and Siuslaw) include any reference to stand area, all of them using 10 acres as the criterion. Only one (Siskiyou) includes any reference to dead trees, and none specifically mentions snags. Shasta-Trinity's definition is broad enough to include not only mature forests but even some stands forty years old. And the Umpqua plan defines old-growth as natural stands of any age, structure, and ecological dynamics. By this definition, a stand of inch-high seedlings is old-growth! No wonder old-growth seems plentiful.

None of these planning documents distinguishes between lowland and subalpine forests, nor among any of the various kinds of old-growth communities. Most definitions are based on subjective criteria (such as someone's idea of "old") rather than objective criteria (such as the density of trees above a specific age). By and large, the definitions are timber-oriented rather than ecologically oriented. And all are inclusive, lumping together very different forest stages, many of which would not be included in an ecologically meaningful definition.

Diverse, vague, subjective, largely timber-oriented definitions complicate efforts to determine how much old-growth remains. And by being too inclusive, they overestimate the amount of ancient forest.

Fortunately, the Forest Service can do better. The most comprehensive published attempt to define old-growth ecologically is the set of interim definitions published by the Forest Service–BLM Old-Growth Definition Task Group (1986) (see Table 3.2). They use multiple criteria rather than a single one such as size or age of trees. They are minimum criteria so that it is not necessary to decide subjectively how they deviate from an "average" value; with a minimum criterion, a stand is either old-growth or is not. Most of them are objective and quantitative, lessening the need for interpretation. As a result, a timber cruiser or citizen conservationist can determine unambiguously whether a stand is old-growth. The definitions are flexible, using criteria tailored to various major Westside forest community types. And they strike a reasonable balance between being too inclusive (which would take in forests that do not really have old-growth character) and too exclusive (which would define most old-growth out of existence).

The interim definitions do not provide all the answers and will need to be improved as new information becomes available. They are not specifically designed for stands dominated by Sitka spruce, western redcedar, western hemlock, or Pacific silver fir, although criteria for Douglas-fir/western hemlock stands apply to them. They do not include old-growth characteristics such as broken tops or epiphytes, which are difficult to quantify. These are hardly serious shortcomings.

Although the authors discussed the need to define a minimum area that could be considered an old-growth stand, they did not settle on one, either the oft-mentioned 80-acre figure or any other. This is an important omission because a small stand—ten, eighty, perhaps any less than several hundred acres—is so subject to external forces that it has little

Table 3.2

INTERIM MINIMUM STANDARDS FOR WESTSIDE OLD-GROWTH

	Douglas-fir on western hemlock sites (western hemlock, Pacific silver fir)	Douglas-fir on mixed-conifer sites (white fir, Douglas-fir)	Douglas-fir on mixed-evergreen sites (tanoak, Douglas-fir)
Live trees	Two or more species with wide range of ages and tree sizes	Two or more species with wide age range and full range of tree sizes	Douglas-fir and evergreen hardwood (tanoak, Pacific madrone, and canyon live oak) associates (40 to 60 percent of canopy)
	Douglas-fir ≥ 8 per acre of trees >32 in. diameter or >200 years old	Douglas-fir, ponderosa pine, or sugar pine ≥ 8 per acre of trees >30 in. diameter or >200 years old	Douglas-fir or sugar pine ≥ 6 per acre of trees >32 in. diameter or >200 years old
	Tolerant associates (western hemlock, western redcedar, Pacific silver fir, grand fir, or bigleaf maple) ≥ 12 per acre of trees >16 in. diameter	Intermediate and small size classes are typically white fir, Douglas-fir, and incense-cedar, singly or in mixture	Intermediate and small size classes may be evergreen hardwoods or include a component of conifers (e.g., Douglas-fir or white fir)
Canopy	Deep, multilayered canopy	Multilayered canopy	Douglas-fir emergent above evergreen hardwood canopy
Snags	Conifer snags ≥ 4 per acre that are 20 in. diameter and >15 ft. tall	Conifer snags ≥ 1.5 per acre that are >20 in. diameter and >15 ft. tall	Conifer snags ≥ 1.5 per acre that are >20 in. diameter and >15 ft. tall
Logs	Logs ≥ 15 tons per acre including 4 pieces per acre ≥ 24 in. diameter and >50 ft. long	Logs ≥ 10 tons per acre including 2 pieces per acre ≥ 24 in. diameter and >50 ft. long	Logs ≥ 10 tons per acre including 2 pieces per acre ≥ 24 in. diameter and >50 ft. long

SOURCE: Old-Growth Definition Task Group (1986).

chance of remaining viable. The criteria cannot be used for aerial interpretations without modification. This is an important omission because assessing the status of stands from the ground can be expensive and slow. Any comprehensive assessment of forest resources would likely use aerial photographs to some degree.

Finally, the interim (or any) definitions have an inherent risk: that someone might think that any natural stands not meeting them are unimportant and can therefore be eliminated with impunity. Nonetheless, for now, the interim definitions are by far the most useful basis for assessing how much old-growth remains, and where.

The Forest Service's various definitions of old-growth remind me of the classic "good news, bad news" jokes. The good news is that it has ample professional expertise needed to develop flexible, clear, ecologically meaningful definitions. The bad news is that, so far, it has shown no inclination to use them.

IV
The Biological Values of Ancient Forests, Part 1

Biological Diversity

The diversity of life forms, so numerous that we have yet to identify most of them, is the greatest wonder of this planet.
(Edward O. Wilson, ed., *Biodiversity* [1988])

In the 1970s, paleontologists in Texas discovered fossil bones of the largest aerial animal ever known, a flying reptile called *Quetzalcoatlus northropi*. This pterosaur, whose wings spanned perhaps 40 feet, was the king of the skies in the Cretaceous era. Fascinated by this huge creature, in 1980 a talented team of aeronautical engineers decided to create a working half-size scale model, with a grant for $500,000 and sponsorship from the Smithsonian Institution's Air and Space Museum. In May 1986, near Washington, D.C., they demonstrated to the media and a crowd of 300,000 what they had achieved. Seconds after the model was released from the tow line, it crashed to the ground and broke.

I do not recount this tale to mock those who boldly took up the challenge but, rather, to point out that our best minds could not match something that nature did more than 65 million years ago, something that not only flew better but also gathered its own fuel and reproduced itself.

Nature has produced an incredible diversity of such living things, from infinitesimal bacteria to trees weighing nearly as much as the combined human population of Port Orford, Florence, and Oakridge, Oregon. The diversity of life is a miracle, but one that now faces its gravest threat. The Earth is fast losing its biological diversity.

There have been inklings of this for a long time. But it has only been since 1979, when Norman Myers's *The Sinking Ark* was published, that awareness about the rapid loss of species has spread beyond a handful of scientists to the general public and decision-makers.

Although the media pay far more attention to sports, movie stars, stock market swings, and military adventures, most thoughtful people realize that these phenomena are transitory. In contrast, extinction is forever. The Earth will never again know the thunderous wingbeats of a billion passenger pigeons or the fluttering wingbeats of a Xerces blue butterfly. If we do not act quickly, the hushed wingbeats of northern spotted owls will likely vanish as well. But the problem goes much farther; the world is now or will shortly be losing living things at a rate without precedent since the disaster that eliminated the dinosaurs 65 million years ago. We were not around at the last mass extinction, but as the cause of nearly all recent extinctions, we have the power to stop this one.

Why should we? In a world where terrorists, illegal drugs, AIDS, and mortgage payments dominate our concerns, why should we burden ourselves with yet another? Why does it matter whether bugs, fishes, pandas, or rainforests continue to exist?

There are really two answers. One, the ethical argument, is persuasive for many people but, regrettably, not, as yet, for most. Excellent discussions appear in David Ehrenfeld's *The Arrogance of Humanism* (1978), Bryan Norton's *The Preservation of Species* (1986), Ed Wilson's *Biodiversity* (1988), and Bill Devall and George Sessions's *Deep Ecology* (1985).

The second answer, however, is utilitarian, based solely on our self-interest. The living things of this Earth are our resources and life-support systems. Every bite of food you eat comes from living things. Every sip of water you drink has been cleansed by living things. The oxygen in your every breath comes from living things. Humankind is utterly dependent on biological diversity.

And it is not just individual organisms and species that we depend on; it is whole ecosystems. The organized study of the interactions of living things and their environment—the science of ecology—was born about the same time that Euro-Americans settled the Westside and came into its own only in this century. The vast majority of relationships among the living things are as yet unexplored. But everywhere we look, ecologists see miraculously intricate interactions that affect humans. Every ecologist has favorite examples. The following is one of mine.

Some years ago, the University of Wisconsin's Stanley Temple (1977) noted that *Calvaria major*, an endemic tree that had once covered the island of Mauritius in the Indian Ocean, was almost extinct, with only thirteen remaining. Further, although the *Calvarias* bore fruit, all were 300–400 years old. Of course, many trees valued for timber have been decimated. But why was there no young *Calvaria?*

Until the age of European exploration, Mauritius had no native people but was home to a gigantic (more than turkey-sized) flightless pigeon, the dodo, whose name has become synonymous for both stupidity and extinction. Dodos, having evolved without mammalian predators, seemed stupid because they neither fought nor ran away when Portuguese and Dutch sailors first encountered them in the 1500s. By all accounts, dodos were nearly inedible, but the sailors made great sport of clubbing them to death. The last dodo was gone by 1680.

Temple noted that this date coincided with the age of the *Calvaria* trees. On a hunch, he took *Calvaria* seeds and, lacking a dodo, force-fed them to the next best thing: a turkey. The seeds passed through the turkey's gizzard, and, when planted, germinated, the first *Calvaria* seedlings in more than three centuries! The great trees had all but disappeared because the dispersal agent needed to abrade their seed coats had been exterminated. Living things are connected in ways that we have scarcely begun to understand.

Biological diversity is the diversity of life. It is useful to envision three levels, the most familiar being the middle one, species diversity, the wealth of species.

Until 1982, systematic biologists had described and catalogued some 1.5 million species of plants, animals, and microorganisms, but they knew that twice, perhaps even six times, as many had not yet been discovered. Then Terry Erwin (1984) of the Smithsonian Institution used an innovative technique to study an almost unexplored realm: the canopies of tropical forests. He found so many undiscovered arthropods that he estimated an astounding total of at least 30 million species. Although people are most familiar with large species—redwoods and wrens, whales and wildflowers—most living things are actually small invertebrates. It took until the 1980s for science to realize the wealth of life on Earth.

Scientific knowledge is incomplete in Westside forests as well. We haven't a clue how many kinds of fungi or mites there are. This is no small shortcoming. These little-known creatures are crucial to ecological processes in their habitats and thus to our lives.

Genetic diversity and ecosystem diversity—the levels of biological diversity below and above species diversity—are even less known but no less important. Within species, individual differences in form, coloration, physiology, and behavior largely reflect differing genes.

Genes are segments of the complex molecule DNA, which codes for the production of proteins in living things. The sequence of gene locations (or loci) and the different versions (alleles) at each locus determine precisely which proteins are produced and when. For example, genes code for production of hormones that stimulate division in certain bone cells on a buck's skull in spring, initiating the formation of antlers. Genes are passed between generations when organisms reproduce, whether sexually or asexually. Thus, they control the heritable characteristics of organisms.

Each species is made up of one or more populations of interbreeding individuals. Within each population, individuals vary genetically. Different populations can have the same collection of alleles. For example, an allele coding for early flowering might be common in some populations of Pacific rhododendrons and rare in others. Or populations can possess alleles that are absent from others. For example, one population of Port Orford cedars might have an allele that confers resistance to nickel, a toxic element in certain soils, while another population might not.

These genetic differences allow organisms to adapt to their particular environments. For example, Douglas-firs west and east of the Cascades have genetic differences that determine such crucial traits as growth rates, resistance to particular diseases and light absorbance. Indeed, they are different enough to be considered distinct subspecies. But genetic differences occur on much more local scales as well. By providing the raw material for evolution, genetic diversity allows populations to adapt to the changing conditions of our heterogeneous world. The ability to adapt becomes increasingly important as we alter the environment in unprecedented ways, with unprecedented speed.

Ecosystem diversity is a higher level of biological diversity. Our planet's countless combinations of climatic, geological, and chemical conditions favor communities of organisms with different needs. These communities in their physical settings are called ecosystems. Ecosystems can be vast: The Westside biogeographic region is an ecosystem, as are North America and, indeed, the entire biosphere. An ecosystem can also be as small as a stream drainage, a stand of trees, or the soil beneath a single downed log.

Ecosystem diversity can be easy to see, as when coniferous forests on Cascades slopes give way to Willamette Valley croplands, grasslands, and oak woodlands. Boundaries between ecosystems can also be subtle and hard to define precisely. But whether or not people can see them, they are real. For example, the Forest Service and the BLM are in the process of designating forest patches as habitat areas for spotted owls. Unfortunately, the owls occupy some stands but ignore others that look similar to people. The owls distinguish among ecosystems using criteria that we do not yet understand.

The scarcity of severe weather, the north-south temperature and precipitation gradients, the east-west fog and precipitation gradients, the rugged topography, the complex pattern of soils, and the rich variety of disturbance regimes combine to make the Northwest a region of exceptional ecosystem diversity.

Differences among populations, species, and ecosystems are important in countless ways. Recognizing this, Congress passed laws including the Endangered Species Act (ESA) of 1973 and the National Forest Management Act (NFMA) of 1976. The ESA requires the federal government to identify populations and species that are endangered (at risk of extinction through all or part of their ranges) or threatened (likely to become endangered), to take no actions that would jeopardize their continued existence, and to act to bring about their recovery. NFMA regulations require the Forest Service to maintain viable populations of native vertebrate species, to provide for plant and animal communities, and to aid recovery of endangered species in our national forests. When agencies adhere to them, these laws provide a useful mandate for conserving biological diversity.

Nevertheless, this mandate is not always fulfilled. Part of the reason is legislative: In drafting the NFMA in 1976, Congress neglected to give the Forest Service an explicit mandate to maintain genetic diversity. But a more important reason is the Forest Service's prevalent pro-timber bias, which has led Forest Service officials to claim that logging increases biological diversity. But as The Wilderness Society's David Wilcove explains:

If one were to census the birds breeding within a large expanse of old-growth Douglas-fir forest in the Pacific Northwest, one would find a variety of forest-dwelling species, including Vaux's swift, hermit warbler, and perhaps spotted owl. Absent would be most of the species associated with weedy fields, pastures, and other disturbed landscapes—such as dark-eyed junco and brown-headed cowbird. By clearing a portion of the old-growth forest, one might be able to attract some of these open-country species. On that particular tract of land, species richness would be increased. However, this increase should not be confused with an increase in biological diversity. Quite simply, there is no shortage of habitat for dark-eyed juncos and brown-headed cowbirds in the Pacific Northwest, whereas the species associated with old-growth coniferous forests are diminishing as the old-growth forests themselves are logged. (1988, p. 3)

Somehow, natural processes created a rich diversity in the Northwest long before logging began. To maintain the remaining biological diversity, forestlands must be managed with a regional, national, or global perspective. Ancient forests are exceptionally important to the biological diversity of the Northwest, the United States, and the world, as we will see.

Species Diversity

ANCIENT FORESTS AS WILDLIFE HABITAT

Nature's stern discipline enjoins mutual help at least as often as warfare. The fittest may also be the gentlest.
(Theodosius Dobzhansky, Mankind Evolving [1962])

If you ask a knowledgeable Northwesterner about the species that make old-growth forests so special, he or she might mention Douglas-firs, spotted owls, sword ferns, chinook salmon, Olympic salamanders, chanterelle mushrooms, Roosevelt elk, Pacific rhododendrons, banana slugs, cougars, or western redcedars. But the list of known old-growth species runs into the thousands, and the real number is undoubtedly a good many thousands.

Some of these species—such as grisette mushrooms and wood ducks—are widespread across North America. Others—including varied thrushes and Oregon-grapes—occur mainly in the Pacific Northwest but are common in some other ecosystems within the region. But for still others, old-growth is especially important.

Although ancient forest ecosystems are important for many reasons, the controversy about their fate has focused largely on their importance as habitat for species (particularly spotted owls) versus their importance to the timber industry. Unfortunately, scientific knowledge about old-growth as habitat has not yet influenced opinions expressed in some newspaper editorials, commercials, public hearing records, political speeches, living rooms, and barrooms. These opinions are fueled more by passion (or fear) than understanding, which greatly complicates efforts to resolve the controversy.

First, a definition: "Habitat" is where an individual, population, or species lives. It is not a synonym for community or ecosystem. Thus, we can speak of devil's club, coho salmon, or black-tailed deer habitat but not riparian, stream, or ancient forest habitat. The boundaries of ecosystems and habitats are not necessarily the same.

For example, Westside rainbow trout occur not only in medium-sized streams but also in rivers, small streams, lakes, and coastal marine waters, although not in all of them. Natural rainbow populations vary genetically in these different ecosystems, and so rainbows vary in their habitat relations. Steelhead and kamloops rainbows, populations in the Klamaths and in the north Cascades, all differ in their habitat relations. There are also variations within populations. Adults are ecologically different from hatchlings or juveniles. Individuals occur in different places in different seasons. Finally, the size and structure of rainbow populations can vary markedly between years of usual and unusual weather.

In other words, the habitats of these fish defy simple characterization. The same is true of grand firs, goshawks, marbled murrelets, spotted owls, red tree voles, and fishers because each has unique requirements and unique ways of meeting them. Moreover, detailed investigations of species/habitat relationships have not even begun for most Westside species, so decision-makers often lack information that they need as pressures on species increase.

This does not mean that we are totally ignorant about ancient forests as habitat. There have been hundreds of person-years of studies by scientists in universities, federal and state agencies, and the timber industry. Added to knowledge from long-standing fields such as zoology, botany, mycology, ecology, wildlife biology, and fisheries biology are exciting insights about species-habitat relationships emerging from the new, synthetic disciplines of conservation biology and landscape ecology. Although no one can guarantee that spotted owls will not slip into extinction during the next century, we know enough to improve the odds of their survival. Whether decision-makers will act to save them is another question.

What features make ancient Westside forests so important as habitat? They seem preeminent in two ways—having the world's highest diversity of giant conifers and the world's greatest accumulations of snags and downed logs—and are not far behind in several other important ecological measures. But their value as habitat comes not from one feature but from their unique combination of features. Because species as different as ravens and mycorrhizal fungi experience the same forests in very different ways, habitat features that are important to one can be irrelevant to another. Some of them are the following.

1. *An exceptionally moist, moderate temperate climate with summer drought.* Total precipitation is exceptionally high, the growing season is exceptionally long, hard freezes and high summer temperatures are rare, winter-summer and day-night temperature differences are small, cloud cover is frequent during much of the year, fog is common, tornadoes and hurricanes are unknown, and thunderstorms are uncommon. That is, the climate is exceptionally moderate; sharp fluctuations and violent weather are rare. The climate within old-growth forests (lower wind speeds, higher relative humidities, slower temperature changes, lesser snow depths) is even *more* moderate, favoring organisms that cannot tolerate severe climates.

For example, in eastern temperate forests, about 75 percent of bird species must leave during the winter, when energy costs of staying warm are high and insects are scarce. But only about 25 percent of the birds of Westside ancient forests winter farther south; the rest reside year-round. The mild climate and year-round availability of food resources are important reasons that animal diversity is unusually high.

2. *Exceptional topographic and climatic diversity.* An eastward transect through the Westside can take you from cool, foggy coastal lowlands and montane rainforest through a valley with hot, dry summers and then into foothill, montane and subalpine forests, alpine tundra, and up to the perpetually frozen nival zone of the tallest mountaintops. Along this transect, there is not one major gradient, but three, producing exceptional physical diversity. There is the coastal-inland gradient, from very moderate to somewhat less moderate climates. There are lowland-montane gradients, along which temperature decreases and precipitation increases. And although the transect extends west-east, there is also a north-south gradient apparent as Sierran species (such as

sugar pine) marching northward in the warm lowlands and boreal species (such as Alaska-cedar) ranging southward along the cool ridgetops. These gradients are especially pronounced in southwestern Oregon and northwestern California, where the biota is exceptionally diverse even by Westside standards.

Further, because the Westside is geologically complex and climatically diverse, there are many different kinds of soils compared with most regions this size. As a result, it has an extraordinary diversity of ecosystems.

3. *An exceptionally diverse disturbance regime.* Fires, windstorms, insect outbreaks, diseases, floods, landslides, and volcanic eruptions cause disturbances with widely varied sizes, frequencies, and seasonalities. This patchy disturbance regime allows many different communities and species to persist. Fire, for example, favors Douglas-firs over western redcedars, whereas laminated root rot has the opposite effect. Ecosystems with more uniform and more severe disturbance regimes have successional mosaics with fewer kinds of patches. For example, coniferous forests in interior Alaska burn so often that old-growth forests could not develop even if the trees had the potential for long life spans. Similarly, in eastern forests, hurricanes and tornadoes appear to prevent trees from reaching great age.

4. *Unequaled diversity of giant conifers.* No place on Earth has more species of conifers that can reach 200 feet and attain such large diameters. Most Westside ancient forests host at least three giant conifer species, and as many as eight can be found within a few miles, along with many other smaller ones. Broadleaf trees are less diverse than in most temperate forest regions, but there are exceptionally tall species of some genera (*Populus, Alnus, Cornus, Arbutus*). The high diversity and exceptional size create exceptionally diverse foods (e.g., seeds) and spaces (such as widely spaced trees, deeply furrowed bark, and large cavities) compared with other coniferous forests.

5. *Exceptional longevity.* The dominant conifers in these forests are exceptionally longevous. The shortest-lived can outlast any southern Appalachian tree except for eastern hemlock. Only coast redwoods form longer-lived closed forests. Long-lived trees provide habitat for weak dispersers that might not reach habitats that appear and disappear quickly, and they allow time for development of slow-growing species, such as lichens and yews.

6. *Exceptional biomass of living plants, snags, and downed logs.* Westside old-growth forests support more living plant material than

any ecosystem thus far measured except coast redwood forests. Further, the combination of very large size and slow decay allows accumulations of standing snags and downed logs that might be unequaled. As a result, species that need dead wood—decomposer fungi and some plants—are exceptionally abundant and diverse. An unusually small fraction of plant production is grazed. Instead, most plant material decomposes first, forming the basis of detrital food webs. This has encouraged species that feed on decomposers, such as red-backed voles (the most specialized fungivorous mammals in North America) and legions of fungus-feeding invertebrates.

7. *Exceptional vertical and horizontal spatial complexity.* The dominant trees in ancient forests have very deep crowns: Branches and foliage extend a long way downward from the tops. Various shade-tolerant trees in the understory add further vertical structure. Because dominant trees have limited ability to extend branches into gaps, as big trees die, the canopy becomes uneven and sunlight streams through the gaps. The well-lighted patches on the forest floor develop diverse shrub and herb layers. Owing to the large crown diameters and gaps in the canopy, distances between dominant trees are greater and more variable than in closed coniferous forests elsewhere, including nearby second-growth stands. Thus, species that have trouble running or flying through dense second-growth can move easily through ancient forests.

Snags and downed logs dramatically increase structural diversity, providing varied microhabitats for species that need hiding places from predators and moist refuges during summer droughts. And large downed logs have an especially important structural role in streams as long-lived baffles and dams. By slowing and diverting streamflow, logs form pools and allow coarse gravels and sands to accumulate. This creates stable aquatic habitats with widely ranging flow rates and sediment sizes, greatly adding to habitat diversity in streams. As on land, the exceptional structural complexity in old-growth streams creates opportunities for species with diverse habitat needs.

How many of these species need ancient forests? To answer this question, we have to clarify "need," something the Forest Service's Len Ruggiero and his coauthors (1988) have done quite effectively.

The highest degree of dependence would be found in species or populations that are endemic to the Westside and occur only in ancient forests. Equally dependent would be a second group, species, or populations that live in other ecosystems at some times but must inhabit Westside old-growth during some particular time, such as when they breed. A third group consists of species or populations that maintain viable populations or produce surplus young only in old-growth and cannot persist in other ecosystems without immigration of individuals from old-growth. Large numbers of individuals might well occur outside ancient forests during times of modal conditions, but their populations would not remain viable in unusual conditions (e.g., drought, severe winter). All three groups would disappear if their Westside old-growth habitats were to be reduced beyond a certain threshold value. A fourth group is less dependent: species for which ancient forests are especially good or optimum habitat, but which maintain viable populations outside Westside ancient forests.

One issue that has surfaced during debate on old-growth species is dependence versus preference. Advocates of increased logging have alleged that certain species do not *depend* on old-growth; they just *prefer* it, implying that the species are engaging in unnecessary luxuriating that could be forsaken if they lived a leaner existence.

Such opinions suggest unfamiliarity with advances in evolutionary ecology in the last half-century. Organisms generally behave the way they do because such behaviors maximize what biologists call "fitness." Fitter organisms may not fly faster, fight more fiercely, or do more push-ups. Rather, they pass on more of their genes.

The ideal measurement of an organism's fitness is something like the number of its great-great-great (etc.) grandchildren. Measuring its production of young, eggs, seeds, or spores is less useful because some progeny—perhaps even all—can fail to reproduce; as a general rule, a fairly small portion of potentially breeding individuals produce a disproportionate share of progeny. Measuring acquisition of resources—food, shelter, or mates—is even less useful. But because fitness can be difficult or impossible to measure directly, such indirect measures are often the best ones we've got.

To be sure, not all behaviors increase fitness: Deer that freeze in the headlights of oncoming cars are less likely to pass on their genes. But unless there is good evidence and theory to the contrary, it is safest to interpret behaviors as means of maximizing fitness.

Some organisms have preference behaviors: They actively choose one habitat over another. Others are more abundant in some habitats than others not because they "prefer" them but because they simply are more successful at surviving or reproducing there. A red crossbill can choose whether to alight in a particular stand, but a redcedar cannot prefer one habitat over another. Wherever a seed lands, it either lives or it doesn't.

But with or without habitat preferences, an organism has higher fitness in some habitats than in others.

So, when an organism's fitness cannot be measured directly, its habitat selection behaviors or abundance patterns provide valuable clues to its requirements. They are not infallible indicators. But when a species consistently occurs, or has consistently high populations, in a potential habitat out of proportion to that habitat's availability, it is probably telling us something.

Spotted owls, for example, do not occur in ancient forests without reason; they live where they have the best chance of passing on their genes. A handful of Westside spotted owls can be found in younger forests. Perhaps a few even attempt to nest in them. But spotted owls occur most and fledge the most young in old-growth.

Similarly, they do not occupy large territories without reason; defending a territory is so costly that it's counterproductive to hold one larger than necessary. But only large territories can ensure owls the resources they need to survive and reproduce. With less habitat, they will produce fewer viable young, if not in an average year, then in a poor year.

So, if spotted owls, by their behaviors, tell us that they need 3,000 acres, it won't do for land managers to "grant" them 2,700. Cheating them out of habitat that eons of evolution have compelled them to seek condemns them to extinction. As Jack Thomas and coauthors state:

> Our knowledge and understanding of old-growth communities is not adequate to support management of remaining old growth on criteria that provide *minimum* habitat areas to sustain *minimum* viable populations of one or several species. The potential consequences and the distinct possibility of being wrong are too great to make such strategies defensible in the ecological sense. (1988, pp. 257–58)

Below, I will examine relationships of some representative species and their ancient forest homes. It is hardly an exhaustive list. There is amazingly little information on habitat needs of even well-studied species, and much that we once thought to be true about species such as deer has turned out to be incorrect.

Virtually no information is available for some groups because research funding to study them has long been very scarce. For example, systematists told me that inadequate information hampers discussion of old-growth dependence among invertebrates such as snails, spiders, and insects. Lest anyone snicker that these small, creepy things are

Elliott A. Norse

Banana slug, Olympic National Forest (Washington). *Invertebrates, such as snails, insects, and spiders, are far more diverse and abundant than vertebrates in ancient forests, and more important ecologically. Unfortunately there has been very little research on their ecological distribution and their roles in Westside forest ecosystems.*

eminently ignorable (compared, say, with "real" wildlife), I should point out that they are vastly more abundant and play larger roles in their ecosystems. Our anthropocentrism, which causes us to identify with our large, warm-blooded relatives, should not lead us to assume that small creatures are unimportant.

In the following sections, three themes recur like leitmotifs. One is that *ancient forests are crucial refuges during hard times.* Their high structural and species diversity serves as a buffer against drought, holocaust, flood, storm, epidemic, and food shortage. Many species can probably use younger forests under benign conditions (when sane biologists do their fieldwork) but would vanish under severe conditions without the refuges that ancient forests provide. Not surprisingly, many species that occur in younger forests nonetheless reach peak abundance in ancient forests.

The second theme is that *eliminating old-growth actually has some benefits in the short term, but its costs far outweigh the benefits for the much longer period when the canopy is closed in the second-growth that replaces it.* This is a dangerous trap, for the immediate benefits are obvious, but the costs only become evident after a few decades. Some people, not realizing this, assume that there are no long-term costs. Others have no inhibitions about reaping whatever benefits they can and leaving any costs to the rest of us.

The last of the three themes is *the remarkable web of interactions among ancient forest species.* Because our minds are better at focusing on one thing at a time than at contemplating something as complex and unfamiliar as a forest ecosystem, people often think of species in isolation, as when officials try to manage Douglas-firs, elk, or spotted owls. But we cannot really manage a species without understanding how it is embedded in its community of organisms, and we cannot manage a community unless we understand how the physical factors in the ecosystem shape it and how the organisms, in turn, alter those factors.

Moreover, our adaptations give us a very biased view of ecological processes. The interactions we know best are those occurring within a few feet above the ground. We know much less about what goes on within the soil below our feet and the canopy above our heads. Interactions in these realms have great impact on the realm we know best.

Scientists now have some idea of what happened to forests on Mauritius when dodos were lost. What would happen if the Westside were to lose Roosevelt elk (which almost happened early in this century), the colossal fungi known only as *Oxyporus nobilissimus* (which might already be extinct), or spotted owls (which the Forest Service admits are likely candidates for extinction in the coming century)?

Sadly, the answer is, "We don't know." True, we have some ideas of species' interrelations. In recent years researchers have begun to see how the owls interact with red-backed voles and flying squirrels, how these rodents interact with underground mycorrhizal fungi, and how these fungi interact with the trees of Westside forests. But people are so ill-attuned to the connections among living things that some species could disappear without our noticing. In many more cases, we would feel the consequences but would never know their cause.

The web of life connects all species. Some of their interactions are direct, some indirect; some are obvious, some cryptic. But when one component of a system is lost, consequences ramify through the forest, the region, the entire biosphere. As the following sections show, despite our dependence on these species and the ecosystems they comprise, the study of their workings and its value to us is only in its infancy. We are only beginning to understand ancient forests.

NORTHERN SPOTTED OWLS

Of all the horrid, hideous notes of woe,
Sadder than owl songs or the midnight blast,
Is that portentous phrase, "I told you so."
(George Noel Gordon, Lord Byron, *Don Juan,* Canto XIV [1823])

When Euro-Americans spread across the United States, they systematically killed the large predators. Grizzly bears, cougars, and timber wolves were eliminated in most of their former ranges. Now that the species whose teeth we fear are gone, the ones in greatest peril are those whose homes we crave. No species better illustrates this than the northern spotted owl, a species in jeopardy because it lives in ancient forests.

I asked my colleague, ecologist/ornithologist David Wilcove, to examine the ecology of the spotted owl.

Of Owls and Ancient Forests

□ *David S. Wilcove, The Wilderness Society, Washington, D.C.*

Two decades ago, the spotted owl was one of the least known, least studied birds in North America. Today it is the subject of intensive study, rancorous debate, and legal battles. This sudden notoriety stems from the owl's specialized habitat requirements in the western portion of its range, where it inhabits old-growth conifer forests. Because old-growth forests on private lands are essentially gone, almost all of the remaining habitat for the spotted owl in the Pacific Northwest is on federal lands. National forests, for example, account for about 68 percent of the remaining habitat; BLM lands in Oregon add another 15 percent, while national parks account for less than 10 percent.

Both the Forest Service and the BLM are in the process of deciding how much old-growth forest to set aside for the owl, and what they decide will in all likelihood determine its fate. The process has been a painful and controversial one. More than any other species, the spotted owl has become a symbol of the ancient forests, a convenient focal point for the energies and passions of conservationists and loggers alike. It is an almost comical situation—all this attention, all this fury, directed toward a fluffy, twenty-two-ounce bird that likes to spend its days napping and its nights in silent pursuit of rodents scurrying about the forest floor.

There are three subspecies of the spotted owl, distinguished by rather subtle differences in size and coloration. The Mexican spotted owl occurs from southern Colorado and central Utah, south in the higher mountains through Arizona, New Mexico, and extreme western Texas into central Mexico. The California spotted owl is confined to the Sierra Nevada Mountains and the coastal mountains south of San Francisco. The northern spotted owl occurs in southwestern British Columbia, western Washington and Oregon, and northwestern California. Although the Mexican spotted owl is currently under consideration for listing as a "threatened" or "endangered" species under the federal Endangered Species Act, the other two subspecies have caused far greater controversy. The total population of northern and California spotted owls is estimated at about 2,900 pairs, and they are declining. Even more ominously, as tens of thousands of acres of prime spotted owl habitat are clearcut each year, populations of these birds are becoming increasingly fragmented and isolated, magnifying the risk of extinction. The spotted owl is listed as an endangered species by the state of Washington, as a threatened species by Oregon, and as a "species of special concern" by California.

Until recently, virtually nothing was known about the ecology of the spotted owl in the Pacific Northwest. A typical account in a field guide might read something like this: "rare, little known, inhabits forests." Studies in the 1970s by Eric Forsman and colleagues (1984) in Oregon, Gordon Gould (1977) in California, and others showed the owl to be more widespread than previously believed, but they also revealed its strong affinity for ancient forests. For example, 95.5 percent of the sites in Oregon where spotted owls were found between 1969 and 1984 were dominated by old-growth forests or mixed stands of old-growth and mature forest. Typical spotted owl habitat in the Westside consists of mid- to low-elevation virgin forests dominated by Douglas-firs. Such forests have mixed age classes, including trees that are very large and very old (200 years and older) and abundant snags and downed logs.

Northern spotted owls are occasionally found in younger, second-growth stands, a fact that invariably attracts the attention of spokespeople from the timber industry. However, the number of owls that successfully breed and maintain long-term occupancy in second-growth is so small as to be simply irrelevant to the long-term survival of the species. Moreover, most of these younger stands have, on closer inspection, been found to contain remnant old-growth trees. Thus, the evidence that spotted owls require old-growth is overwhelming.

Exactly why they are so closely tied to old-growth is unclear. Several factors are probably involved. First, spotted owls in the Pacific Northwest nest in large, live trees with cavities, broken tops, dwarf mistletoe, or platforms of branches capable of holding organic matter suitable for use as a nest, the sorts of trees one finds in ancient forests but not in younger stands. Second, the prey of spotted owls—mostly small mammals such as flying squirrels, tree voles, and woodrats—may be more abundant in old-growth than in younger stands. Third, spotted owls, like Roosevelt elk or black-tailed deer, use old-growth stands for thermal cover. And, fourth, ancient forests may provide safety from the spotted owl's predators, such as the great horned owl, which often occur in young stands or along the edges of clearcuts.

Not only do spotted owls need old-growth, they need lots of it. For example, studies in northwestern California showed spotted owls using about 1,900 acres of old-growth per pair. Six pairs in Oregon had an average of 2,264 acres of old-growth per home range. Some of these birds were studied for only three to four months. Had they been studied for longer periods of time, their home ranges probably would have increased in size and included larger amounts of old-growth. Six pairs of owls studied in Washington used about 3,800 acres of old-growth per pair, although not all of this old-growth was in one place. During the winter, the Washington owls moved to different forest tracts. Additional data on other pairs, collected within the last few years, confirm these approximate values. Thus, the evidence to date suggests a north-south gradient in the amount of old-growth used per pair of owls. No one is really sure why such a gradient exists.

Critics who ceaselessly argue that more research is needed before any management decisions are made should spend a year or two tracking these nocturnal birds across the rugged terrain of the Northwest. They should also note that further studies have tended to confirm or enlarge the amount of habitat we understand to be necessary for a pair of owls.

Logging of ancient forests is without question the biggest threat to the survival of the spotted owl. Only a small fraction of the original old-growth remains. Also, the remaining old-growth forests are heavily fragmented and now occur in a patchwork of old-growth and clearcuts of various ages. This fragmentation has isolated populations of owls— such as the birds on the Olympic Peninsula—and reduced the probability that juveniles can disperse successfully into unoccupied patches of old-growth.

Fragmentation of old-growth may also be playing a role in two other threats to the spotted owl: competitive displacement by barred owls and predation by great horned owls. The barred owl, a close relative of the spotted owl, has been expanding its range into the Pacific Northwest over the past two decades. No one is exactly sure what is causing this range expansion, but modern forestry may be at least partly responsible. Regenerating clearcuts often have a higher deciduous component than old-growth stands, and this seems to suit the barred owl. Moreover, barred owls do well in second-growth stands, whereas spotted owls do not. Because barred owls are slightly larger and more aggressive than spotted owls, they seem able to displace spotted owls from suitable habitat. Barred owls also seem to be better dispersers than spotted owls.

Great horned owls prey on young spotted owls. They are also quite tolerant of the edges and openings created by logging operations and may use them to infiltrate spotted owl habitat. Spotted owls, in turn, expose themselves to predation when they fly near or across clearcuts during their evening perambulations, something they are being forced to do with increasing frequency these days. How much of a threat the great horned owl poses for the spotted owl is unknown because no one has studied great horned owl predation under different scenarios of forest fragmentation.

Finally, there is the unpredictable but ever-present threat of natural catastrophes. The eruption of Mount St. Helens in 1980 eliminated about 25,000 acres of forest known to contain spotted owls. The recent Westside fires have also consumed many acres of prime spotted owl habitat. Of course, spotted owls have coexisted with volcanic eruptions, fires, windstorms, and other natural disasters for millennia. But that was when there was far more old-growth than exists today and when spotted owls were far more numerous. Populations of spotted owls today are more vulnerable to natural disasters than before extensive logging of Westside forests began.

Besides the obvious problems of habitat loss and fragmentation, what is the evidence that spotted owls are in trouble? The answer lies in the extensive data that have been gathered on the demography of this species. Most individuals do not breed until they are three years old, a surprisingly late start for this medium-sized owl. Reproduction by spotted owls also appears to fluctuate dramatically and unpredictably from year to year. In some years most pairs in a given area will breed, while in other years few even attempt to nest. Some researchers have suggested that this variation in breeding success is due to fluctuations in prey abundance, but this is only a hypothesis: No one has studied the prey base in sufficient detail to confirm or refute the idea.

How, then, do we interpret events in the state of Washington, where the majority of spotted owls have not had a successful breeding season since 1983? Is this part of a normal cycle? Or is it evidence of a chronic decline caused by loss of habitat?

Juvenile mortality of spotted owls has been extraordinarily high, both before and during dispersal. From 1982 to 1984, thirty-one young owls were radio-tagged in Oregon; none survived as long as two years. Since studies began in the early 1970s, the overall first-year survival of young spotted owls has averaged just 11 percent. These data, when applied to standard life table analyses or more complicated models, point to a declining population.

Since the early 1970s, the Forest Service and the BLM have been grappling with the issue of spotted owl management, often under legal pressure from conservation organizations. It has been a byzantine process, the full details of which can be found elsewhere. In this essay I highlight only the more recent milestones.

The Forest Service's mandate with respect to the spotted owl is clear: It is required by law to ensure viable populations of all native vertebrate species occurring in the national forests. In the case of the spotted owl, the Forest Service has tried to fulfill its mandate by creating a matrix or network of old-growth habitat areas. The challenge, of course, is to create a matrix of old-growth areas that is essentially self-sustaining. If the habitat areas are too few in number or too far apart, the owls will have difficulty dispersing between them, and the existing populations within these fragments will wink out, one after the other. Similarly, if individual habitat areas are too small to sustain a breeding pair of spotted owls or if the habitat within them is of marginal quality, the birds are unlikely to survive.

In 1984, the Forest Service released its "Regional Guide for the Pacific Northwest," which contained what were to be final guidelines for spotted owl management in Oregon and Washington. The agency planned to set aside 1,000 acres of old-growth for each of 263 pairs of owls. A coalition of conservation organizations quickly appealed the regional guide, charging that the agency failed to provide an adequate analysis of the likely impacts on the owl and its habitat. The appeal was successful, and the Secretary of Agriculture's office instructed the Forest Service to reexamine its spotted owl guidelines and to prepare a supplemental environmental impact statement (EIS).

In July 1986, a draft EIS was released for public comment, and public comment it got—over 40,000 letters, postcards, and petitions. Most were form letters and postcards generated by the timber industry in opposition to any plans to protect owl habitat. The Forest Service spent an additional two years revising the EIS before issuing a final document in August 1988.

The final EIS calls for the creation of a well-distributed network of habitat areas for spotted owls in Washington and Oregon. The sizes of the habitat areas would vary, depending upon location. In the Klamath Mountains, each habitat area would contain 1,000 acres of suitable habitat. Values for the Oregon Cascades and the Oregon Coast Range are 1,500 and 2,000 acres, respectively. In Washington, habitat areas will contain 2,200 acres in the Cascades and 3,000 acres on the Olympic Peninsula.

For many scientists and conservationists, the final decision was a disappointment. All of the acreage figures fall well below what empirical studies show the owls are using (recall that spotted owls use an average of 2,300 acres of ancient forest per pair in Oregon and 3,800 acres in Washington). Nor does the Forest Service intend to protect very much of the owl's habitat where doing so conflicts with timber production. Only 14 percent of the total number of spotted owl habitats on Forest Service lands designated suitable for timber production will receive protection.

Under this plan, the Forest Service projects only a moderate to low probability of persistence for spotted owl populations after 100 years. A moderate probability provides for "no latitude for catastrophic events affecting the population or for biological findings that the population is more susceptible to demographic or genetic [extinction] factors" than is currently assumed. A low probability means that "[c]atastrophic, demographic, or genetic factors are likely to cause elimination of the species from parts or all of its geographic range during the period assessed" (USDA Forest Service, 1988, p. IV-34). For these reasons, conservationists have challenged the agency's spotted owl plan in court.

The situation is even worse on the BLM lands. The BLM's policy has been to protect the spotted owl only where protection does not interfere with commercial timber harvesting. Since 1977, the agency has set harvest restrictions on 110 spotted owl sites, effective through 1990. The intent is to protect linkages and habitat for ninety pairs of spotted owls between Forest Service lands in the Oregon Cascades and the Coast Ranges and to preserve the integrity of these sites for the next planning period. BLM habitat areas for breeding pairs of spotted owls contain only 300 acres of old-growth each, an amount far less than what studies show that the owls need.

In early 1986, the BLM was directed by the Secretary of the Interior's office to review the status of the spotted owl on its lands. The bureau then appointed a six-member analysis team. In its report, the team concluded that continued harvesting of old-growth on BLM lands would limit the agency's ability to provide more than 300 acres of old-growth per pair and would further fragment the habitat of the owls. In the spring of 1987, the BLM announced that it would not reconsider its current timber management plans with respect to the spotted owl until at least 1990. That decision has been challenged in a lawsuit brought by a number of environmental organizations.

Given the rarity of the northern spotted owl and the threats to its existence, people often ask if it is a federally listed endangered or threatened species. In 1987, a number of local and national environmental organizations petitioned the U.S. Fish and Wildlife Service to list the northern spotted owl. After studying the situation, the agency announced in December 1987 that "listing the northern spotted owl as an endangered or threatened species is not warranted at this time." In its press release, the agency stated that "sufficient data are not available to determine with certainty the long-term trend of the spotted owl population."

Once again, scientists and environmental organizations were disappointed. By 1987, more was known about the ecology and population dynamics of the spotted owl than for most North American birds. The loss of habitat, low numbers, high rates of juvenile mortality, and poor reproduction, coupled with the invasion of the barred owl, all seemed to make the northern spotted owl a perfect example of an endangered species. Many observers felt that politics rather than biology was behind the agency's decision. Indeed, an investigation by Congress's General Accounting Office (GAO) concluded that the evaluation of the spotted owl petition had been beset by problems. Inadequate time was allotted to the spotted owl study team to conduct its analysis, and management within the Fish and Wildlife Service changed the body of scientific evidence presented in the study team's status report after it had been reviewed by outside experts. "The revisions," noted the GAO, "had the effect of changing the report from one that emphasized the dangers facing the owl to one that could more easily support denying the listing petition" (1989, p. 1).

A coalition of conservation organizations sued the Fish and Wildlife Service for violating the Endangered Species Act by refusing to list the northern spotted owl. On November 17, 1988, U.S. District Judge Thomas Zilly in Seattle ruled that the Fish and Wildlife Service decision was arbitrary and capricious, contrary to law. He ordered a reanalysis and new decision by May 1, 1989. In late April, the Fish and Wildlife Service released its new analysis and announced that it would propose listing the northern spotted owl as a threatened species. For conservationists, the agency's reversal was a welcome victory. However, conservationists were also quick to point out that the Fish and Wildlife Service's proposal is just that: a proposal. The agency will have as much as a year to study the situation before reaching a final decision, at which point the owl might—or might not—be listed.

As the controversy grows, it is worth remembering what the spotted owl is and isn't. It is just one species in the constellation of plants, animals, fungi, and bacteria that inhabit the ancient forests, and in an ecological sense it may be no more important than any of these other species. But because spotted owls require so much old-growth, protecting them will mean de facto protection for many other species with smaller area requirements. In this respect the spotted owl is quite special. And based on what we know now, it is the most critically endangered denizen of the ancient forest. (Adapted in part from David S. Wilcove, "Public Lands Management and the Fate of the Spotted Owl," *American Birds* 41 [1987]: 361–67.) □

OTHER BIRDS

Because of the great diversity of ecosystem types, western Oregon supports more bird families than any other area in North America. (Larry D. Harris, The Fragmented Forest [1984])

If you visit the coast and know precisely what to look for, where and when, you might find a clue to one of the Northwest's most enduring mysteries. Within a mile of shore, you might spot a chunky, robin-sized, thin-billed, brown seabird. If not flying low over the swells, it will probably be sitting on the surface, diving and then flying underwater in pursuit of small fishes and shrimplike crustaceans.

Nothing unusual about that; many species in the family Alcidae behave similarly. Rather, it is the bird's nesting habits that distinguish it. To avoid nest raiders, most seabirds nest on small islands lacking mammalian predators or on craggy sea cliffs too steep for predators that cannot fly. Auklets, guillemots, puffins, murres, and most other alcids have chosen this latter strategy. But not marbled murrelets. Wildlife biologist David Marshall (1988) has meticulously summarized the murrelet's unusual breeding biology.

Throughout much of this century where it nested was a mystery. There were clues, true enough, but no conclusive ones. On a continent crammed with so many bird watchers and ornithologists, it is hard to imagine where even a rare bird could nest and avoid being seen. Yet it did. But with patience and luck, you might see a marbled murrelet take off and head inland, not toward towering sea cliffs but toward towering trees in ancient coniferous forests. For this is where marbled murrelets nest.

There are two marbled murrelet subspecies. The Asian one occurs from Japan and Korea to the Soviet Union's Kamchatka Peninsula and Komandorskyi Islands. The American one occurs from the Aleutian Islands to central California in nearshore open coastal waters, bays, sounds, and even in some lakes not far from the coast.

Marbled murrelets differ from other alcids in ways that suggest differences in nesting habits. Unlike others, their limbs are not adapted to burrowing. Murrelets lack the characteristic pungent odor of burrow-nesting seabirds. And their mottled brownish breeding plumage suggests protective coloration that matches moss and wood, unlike the black and white of alcids that nest on sea cliffs.

For decades, people have seen and heard marbled murrelets in or near Westside ancient forests. Some birds were adult females that were collected with eggs in their oviducts, some were adults carrying fish, as if bringing them to their nests. Others were downy chicks or fledglings. And in 1953, a logger found a stunned marbled murrelet and fragments of an egg in the debris of a large hemlock in the Queen Charlotte Islands, British Columbia. But he found no trace of a nest.

In 1961, the first marbled murrelet nest in a tree was discovered in Siberia in a lichen-draped larch several miles from the sea. The first

North American nest was found in 1974, in a depression on a large moss-covered Douglas-fir limb in northern California. Since then, more tree nests have been found in the United States, the USSR, and Japan. More observations have provided needed details. In Alaska, evidence of nests has come from both large conifers and steep tundra or rock cavities where well-developed coniferous forests are absent. All the nests and evidence of nesting from British Columbia southward come from large conifers. Nine of eleven chicks and twenty of thirty-one fledglings have been recorded in what are described as old-growth forests, all the rest near old-growth forests, often in steeply sloping areas. The broad branches with thick layers of epiphytic lichens and mosses where nests have been found occur only in trees of perhaps 150 years and older. From this evidence, ornithologists concluded that marbled murrelets in the Pacific Northwest need old-growth trees in coastal areas.

Hence, murrelets are headed for trouble. Like spotted owls, they are long-lived birds that have far less ability to recover from diminished populations than birds that breed early and lay more than one egg per year. Oil spills and gill-netting threaten them in their marine habitat, and destruction of old-growth forests threatens them in their breeding habitat. Only a small percentage of coastal ancient forests remain, mainly in Olympic, Siuslaw, and Siskiyou national forests, on BLM lands, where old-growth is disappearing very quickly, and in the few larger national and state redwood parks in California. A large and growing portion of coastline has no ancient forest and lacks marbled murrelets during the breeding season.

Because murrelet nesting habitat is rapidly disappearing, scientists in the Pacific Seabird Group issued a resolution of concern in 1986. It called upon state, provincial, and federal agencies to establish an interagency working group to address research and management needs and urged the Forest Service to designate the marbled murrelet as a sensitive species in management plans for national forests within 75 km of saltwater. The Forest Service has not yet done so. In 1988, the National Audubon Society and a number of Northwest conservation organizations proposed marbled murrelets for listing as threatened (a status indicating that it is likely to become endangered if appropriate actions are not taken in the near future) under the U.S. and Oregon Endangered Species acts.

Determining the population status of marbled murrelets will be neither easy nor inexpensive. As murrelet researchers attest, following a small, fast-flying bird as it moves over the surf and through the old-growth canopy in semidarkness poses some problems. Further, radio telemetry does not always work well for marine species.

Nor will efforts to study and conserve marbled murrelets be free of controversy. Some people in the timber industry have derided proposals to protect marbled murrelets as yet another "plot" to "lock up" a resource that they view as theirs: ancient forests on public lands. Even some newspapers have revealed little concern or understanding. For example, a January 20, 1988, editorial in the *Curry Country Reporter* of Gold Beach, Oregon, called the marbled murrelet an "antilogging tool" and implied that the Westside's small populations do not merit protection because larger ones are still found in Alaska. By the same logic, Northwest timber industry jobs should not be protected because there are more industry jobs in the South. The editorial also neglected to mention that the Endangered Species Act requires the federal government to list species that are imperiled throughout all or substantial parts of their ranges. One would think that California, Oregon, and Washington might qualify as "substantial."

The controversy over destruction of spotted owl habitat has so dominated editorial pages, hearing testimony, and governmental wildlife research outlays that habitat relations of other species have largely been overlooked. But marbled murrelets are only one among a number of birds that deserve attention because their fate appears to hinge on the availability of suitable old-growth habitat. Another is Vaux's swift.

Swifts are aptly named small birds that are superbly suited for life in the air, pushing the extremes of avian design. So minute are their feet that swifts can barely crawl. But swifts are remarkable flying machines, with abbreviated tails and downcurved, swept-back, crescentic wings that make the birds look like stealth bombers.

Swifts appear to hold the world's record for continuous flight. The record for an airplane is almost sixty-five days. But some swifts are believed to spend all their time in the air between fledging (leaving the nest for the last time) and building their own nests two to three *years later*! And while performing aerobatics that no airplane can match, they fuel their engines with moths, flies, and mosquitoes.

But not even swifts can do everything in the air. Vaux's swifts nest and roost in large groups, mainly in very large, hollow old-growth snags, often ones that have been burned out. Not surprisingly, Dave Manuwal and Mark Huff (1987) found Vaux's swifts to be far more abundant in ancient forests than in younger forests, suggesting that both stand structure (which affects the foraging for these aerial fly-catching birds) and presence of nesting sites might account for this pattern.

Compared with the more widespread chimney swifts of the eastern United States, Vaux's swifts show much less tendency to nest in chimneys. Because intensively managed forests offer no trees with large, chimneylike cavities, ancient forests are the primary natural habitats that can support their continued existence.

One of the odder sights you might see in an ancient forest is something looking like a mouse running up a large tree trunk. But if your eyes are good, you will notice that the "mouse" has feathers and a slim, decurved bill. This rather unbirdlike songbird is a brown creeper, the only North American species in its family.

Summer is bountiful in the temperate zone. When winter relaxes its grip, many birds migrate from the tropics and subtropics to take advantage of the annual pulse of insects, fruits, and other foods. With ample protein-rich food and diminished energy costs for staying warm, songbirds can afford to produce one or two clutches of young. But throughout much of the temperate zone, when winter returns and these foods disappear, most birds are forced to migrate to regions where they can keep their energy balance "in the black," gaining enough calories from food to cover the losses from stoking their metabolic "fires."

In much of the United States, brown creepers have to migrate. But in the mild Westside climate, brown creepers and, indeed, a large fraction of the avifauna are year-round residents.

As Jina Mariani (1987) of the University of Washington has recently shown, in the southern Washington Cascades, mature and ancient forests appear to be optimal habitats for brown creepers, for they offer both nesting sites and year-round food resources. Brown creepers usually nest in snags where sheets of bark have separated from the wood. Since the interval when bark has partly separated but not completely sloughed off is brief, few snags offer brown creeper nest sites at any one time. Hence, the best nesting habitats are ones that produce good numbers of snags steadily, not only at stand initiation.

Mariani's research illustrates another general point: Different trees have different value as habitat elements. Although some brown creepers nest in Douglas-firs, they greatly prefer western white pine, a once-common species that has been devastated by white pine blister rust. Management that eliminates preferred tree species will be more harmful for these birds than would be predicted from stand age alone.

The importance of nesting habitat has been emphasized in many studies, but survival in nonbreeding seasons can be crucial. Manuwal and Huff (1987) studied bird communities in spring and winter along a forest age gradient in the southern Washington Cascades. They found no major differences in species diversity among different aged stands in spring, although some species (such as Vaux's swifts and brown creepers) are more abundant with increasing stand age.

The picture is very different in winter, when migratory species have left. While there is no marked difference in species richness among stands of different ages, one species (white-winged crossbill) was present only in old-growth, and birds in general were three times more common than in young-growth. For three finch species, this was probably a response to the greater abundance of western hemlocks (and their seeds) in old-growth stands. Unlike Douglas-firs, hemlocks produce cones every year. Manuwal and Huff suggest that differences between natural stands and even-aged tree plantations could be even more pronounced than differences among natural stands of various ages.

ELK AND DEER

In the late 1970s, the population of elk was estimated at approximately 500,000 ... some 10 million elk of six subspecies existed in North America at the time of the arrival of Europeans.
(Jack Ward Thomas and Larry D. Bryant, *Audubon Wildlife Report 1987*)

One of the least-questioned ideas about forest wildlife is that cutting virgin forests is good for members of the deer family. An old idea, its influence is clear even in some recent publications. Its most extreme adherents assert that old-growth is a biological desert with insufficient forage, that deer *need* forest edges and that intensive timber management is good for deer. Federal forestry officials have used these ideas to justify a great deal of logging.

Unfortunately, although logging did benefit deer when the Westside was solid old-growth, in balance, it does not today, especially where heavy snows can occur. A careful examination of the literature by John Schoen, Olof Wallmo, and Matthew Kirchoff (1981) suggests that early observers failed to make the crucial distinction between mature and old-growth forests and that virgin forests are actually better deer habitat than forestlands that have been cut. Bolstering their review is recent research in the Westside and southeast Alaska. Because deer are a major recreational resource and because so much old-growth has been cut (ostensibly) for their benefit, it is important to understand their habitat requirements when discussing the future of ancient forests.

Roosevelt elk are one of the world's largest deer and perhaps the Northwest's most prized game species. Their range coincides precisely with the region covered in this book (except that, like Douglas-fir forests, they cross into Canada). This suggests that their ecology is closely tied to the ancient coniferous forests that distinguish the Westside from the rest of North America.

As Ken Raedeke and Dick Taber (1982) of the University of Washington point out, elk were important foods of many Native American tribes, who were careful not to overhunt them. But Euro-Americans in the 1700s and 1800s viewed elk not as sacred foods or magnificent game animals but as large, free sources of meat. They decimated elk across North America, driving two of the six subspecies (eastern and Merriam elk) to extinction. Roosevelt elk were so rare by 1900 that their hunting was prohibited. With four decades of recovery, populations increased, carefully regulated hunting was allowed again, and hunters now kill more than 3,000 each year in Oregon alone.

Elk are usually thought of as grazers, but Roosevelt elk eat a variety of grasses, herbs, and shrubs, even tree seedlings and foliage. Although all elk need forests, Roosevelt elk seem to depend on forests more than the other subspecies and can reach high population densities in ancient forests. Timber operations have mixed effects on elk. Studies in Oregon cited by James Harper and colleagues (1987) of the Oregon Department of Fish and Wildlife show that cutting old-growth sometimes benefits them in the short term but usually has far larger adverse consequences.

Elk feed more on some species than others because plants differ markedly in their abundance and value as forage. Some are not preferred because they are too low in protein content; others have high concentrations of chemical compounds that deter feeding. Both protein content and chemical defenses of forage plants can vary among seasons. Plants are usually most nutritious and palatable after new growth has begun in spring and least so in winter, when elk need substantial amounts to avoid hypothermia. As a result, food is more likely to be limiting in winter than in the warm part of the year.

In the first year after clearcutting, there is little elk forage, but it rises sharply, reaching higher levels than occur in ancient forests. In a landscape completely blanketed by old-growth, cutting increases elk populations. So if some cutting is good for elk, wouldn't it be better to eliminate all the old-growth?

The answer is no for several reasons. One is that forage always decreases. As conifer canopy coverage approaches 100 percent, grasses, herbs, and shrubs are shaded out and disappear. This occurs fastest in intensely managed tree plantations where herbicides kill forage plants that compete with conifers, and fertilizers hasten canopy closure.

Intensely managed stands have more forage than old-growth for only a few years after cutting. For the rest of the rotation, tree plantations have much less and are much poorer elk habitats. In a couple of centuries, as old trees die, the canopy would begin to open and forage levels would then rise, but intensive forestry does not allow tree plantations to reach this stage. If tree plantations are managed on a 100-year rotation and equal amounts are cut each year, less than 20 percent will have more forage than old-growth at any given time, and at least 80 percent will have less. The tree plantation hare leaps ahead of the old-growth tortoise, only to lag for the rest of the race.

Foresters may not realize that the *quantity* of forage is not the only consideration; two others are forage quality and *availability*. Clearcutting improves forage quantity for a while (if it is not followed by herbicide spraying). As Fred Bunnell and Greg Jones (1984) show, there is conflicting information about its effect on forage quality. But without question, clearcutting can reduce forage availability. Snow accumulations are up to six times deeper in clearcuts than in forests with canopies having 70 percent coverage. Elk avoid clearcuts when they have difficulty plowing through snow and their food is covered. Deep snows are uncommon in the Puget Lowlands and in Willamette Valley, where elk are long gone. But they are common at higher elevations—the world's record yearly snowfall (more than 100 feet!) was at Paradise on Mt. Rainier—and even the Coast Ranges and the Siskiyou and Klamath mountains get snows deep enough to prevent elk from using clearcuts. Indeed, in cutover landscapes, deep snows are especially devastating where they are rare because survivors of mild winters die of starvation and hypothermia in large numbers. Heavy snows—occurring perhaps once a decade—can be more important to elk populations than usual conditions.

Food is not enough. Elk also need cover to escape from harassment and predation (formerly from bears, wolves, and cougars, now mainly from humans) and to avoid temperature extremes. As Harper and his colleagues (1987) explain, optimal Roosevelt elk cover has four layers:

an overstory that can cast deep shade and intercept substantial amounts of snow, and subcanopy, shrub, and herb layers. This structure is characteristic of ancient forests but not of tree plantations. Optimal cover is important both in areas subject to hot weather, such as southwest Oregon and northwest California, and wherever deep snows preclude foraging in clearcuts. Thus, it is not surprising that Roosevelt elk prefer old-growth over younger stands in winter. But as Gary Witmer (1982) found in the Oregon Coast Range, they prefer old-growth during the rest of the year as well.

Although clearcutting old-growth increases the amount of forage for a few years, it also increases the risk of winter starvation from the beginning, eliminates forage for most of the rotation, and decreases hiding and thermal cover. Whereas a landscape consisting of the right mix of clearcuts and closed-canopy tree plantations could provide excellent forage and fair thermal cover, succession makes it impossible to main-

Elliott A. Norse

Columbian black-tailed deer, Olympic National Park (Washington). *Altered landscapes of clearcuts and tree plantations can support large populations of black-tails and Roosevelt elk in years of good weather, but severe snows often cause population crashes. In contrast to this "boom-and-bust" pattern, ancient forests provide the food and cover that deer require in good weather and snowstorm alike, and thus support stable populations.*

tain the correct ratio: Far too much of the landscape will be stages where forage is virtually absent. Further, there is high risk of starvation whenever heavy snows prevent elk from foraging in clearcuts. Harper and his co-researchers at the Oregon Department of Fish and Wildlife have found that extensive clearcutting generally causes elk populations to boom, and then bust. If the goal is to maintain substantial, stable elk populations, eliminating old-growth is the wrong way to do it.

Much the same is true of Columbian black-tailed deer, the Westside's mule deer subspecies, although there are some interesting differences. One concerns the need for thermal cover in different seasons. Adult Roosevelt elk weigh roughly 400–700 pounds, but Columbian black-tails weigh only a quarter as much. For two animals with similar shapes, one with four times as much volume has only about two and a half times as much surface area. Because heat *production* is proportional to volume but heat *loss* is proportional to surface area, elk are at greater risk of overheating in summer, while black-tails are at greater risk of hypothermia in winter. This difference has little effect on forest management, however. The thermal cover provided by old-growth is optimal in both hot and cold weather. But it does mean that black-tails are especially vulnerable during winter.

Black-tails are usually characterized as browsers, but like elk, their diets are broad. Young clearcuts provide them with more forage than ancient forests, but closed-canopy forests provide them with less (Figure 4.1). Thus, it is not surprising that Schoen and Wallmo (1979) found that Sitka black-tails (a closely related subspecies) in southeast Alaska Sitka spruce/western hemlock forests use second-growth only one-fifth as much as old-growth.

Besides losing heat faster, black-tails are less powerful than elk and have shorter legs, so they have more difficulty plowing through deep snow. But it does not take much snow to cause problems: Dick Taber and Thomas Hanley (1979) point out that as little as three or four inches of new snow can drive Columbian black-tails from an area, and they begin to starve quickly when just a few inches cover evergreen trailing blackberries, one of their primary winter food plants.

There is yet another reason that ancient forests can be important to black-tailed deer. Unlike clearcuts and second-growth, they have large amounts of lichens. Lichens live on soil, rocks, and in trees. They dry quickly and all but cease functioning when they are not wet. They lack roots to gather water and nutrients, and their treetop habitats are not exactly rich in minerals. As a result, lichens grow very slowly. Tree-

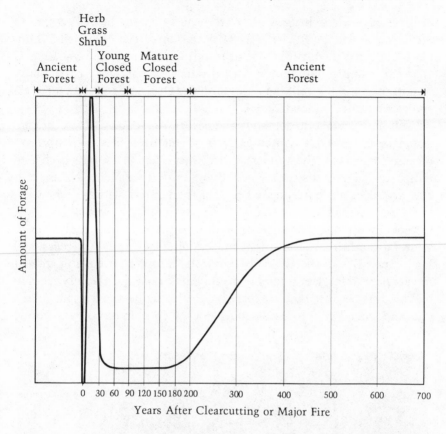

Figure 4–1.   Deer forage during succession in Westside forests. *Food becomes plentiful a few years after clearcutting or a major fire because much light reaches the ground. Then, after the forest canopy closes, food plants decline to a very low level. They remain scarce for several hundred years, until old trees begin to die, creating sunny spots on the forest floor where forage plants prosper for centuries. In managed forestlands, the old-growth stage is eliminated, and most of the landscape is covered with closed-canopy young forest, the stage with the lowest forage levels.*

dwelling lichens do not attain significant biomass until trees are old.

Unlike plants that black-tailed deer eat in winter, lichens are highly digestible and would be an excellent food source if deer could reach them on tree trunks and in the canopy. Since deer are not very good climbers, the lichens must come to them, which is what they do when strong winds blow. After a windy day, the floor of an ancient forest can be littered with lichens, which are eagerly eaten. Indeed, as Susan Stevenson and James Rochelle (1984) of the University of British Columbia found on Vancouver Island, litterfall of lichens (mainly *Alectoria* spp.) can equal production of rooted foliage. Lichens are particularly important for black-tails (and for elk in the Oregon Coast Range) when snows have covered food plants.

With high metabolic demands and decreasing food quality, deer condition deteriorates as winter progresses. But litterfall of *Alectoria* in ancient forests is highest in late winter, just when black-tails need food most. After a winter storm, when high-quality food plants are unavailable and deer are hard-pressed to fuel their internal fires, lichens on snow are manna from heaven.

In deciduous forests, where snow interception and thermal cover in winter are minimal, deer might indeed fare best in early successional landscapes. But in the Westside today, the notion that cutting ancient forests is good for deer is largely myth. Black-tails and elk only seem to prefer edges because they are forced to commute between brushy clearcuts and second-growth stands to replace the combination of year-round

Elliott A. Norse

Lichens (*Alectoria*) on Douglas-fir trunk, Willamette National Forest (Oregon). *Clumps of highly digestible* Alectoria *blown from old-growth trees are important food for black-tailed deer when snow covers forage plants in clearcuts. Lichens are slow-growing and are abundant only in ancient forests.*

forage and cover they had in their natural ancient forest habitats. And intensive timber management does increase forage for a short while when high-quality food is abundant anyway, but it prevents access to forage when food is scarce and eliminates forage for a much longer time.

The Westside is long past the point where additional old-growth cutting would improve black-tail and elk habitat. The best way to maintain substantial, stable deer populations includes protecting the ancient forest that remains.

OTHER MAMMALS

Though only a few inches long, so intense is [the Douglas squirrel's] fiery vigor and restlessness, he stirs every grove with wild life.... Nature has made him master forester and committed most of her coniferous crops to his paws.... [T]he greater portion [of cones] is of course stored away for food to last during the winter and spring, but some of them are tucked separately into loosely covered holes, where some of the seeds germinate and become trees.
(John Muir, "The Douglas Squirrel," *The Mountains of California* [1894])

Deer are the most visible mammals, but Westside ancient forests are also home to an exceptional diversity of smaller, less conspicuous ones. Marty Raphael has been studying mammals of Westside forests for nearly a decade. His research on their distributions in northwestern California forms much of the basis for the following essay.

Mammals of the Ancient Forests

□ *Martin G. Raphael, U.S. Forest Service, Forestry Sciences Laboratory, Rocky Mountain Forest and Range Experiment Station, Laramie, Wyo.*

Mammals constitute about 25 percent of all land vertebrate species in the Westside region of the Pacific Northwest. Because the Westside has so many varied environments, mammalian species diversity is exceptional, second in the United States and Canada only to the Sierra Nevada of California. A total of 110 mammal species occur within western Oregon and Washington; ninety-seven are found in northwestern California. Among all Westside habitats, forests support most of these mammal species, and Douglas-fir forests are especially important. In Oregon and Washington, eighty-two species (75 percent) are associated with Douglas-fir. Similarly, seventy-one species (73 percent) occur in Douglas-fir forests of California.

Not only do Douglas-fir forests support a rich variety of mammal species; they also contain several very distinctive ones. As summarized by Larry Harris and Chris Maser (1984), these include the only exclusively needle-feeding mammal in North America: the red tree vole. Also included is the closely related white-footed vole, which, like the red tree vole, occurs nowhere else in the world. Distinctive species also include the western red-backed vole, the only mammal in North America that feeds predominantly on mushrooms and truffles; the northern flying squirrel, the only lichen-consuming rodent; the mountain beaver, the world's most primitive rodent (also found nowhere else in the world); the shrew-mole, a curious beast found only in the Pacific Northwest, whose nearest relatives occur in Japan and China; and the forest deer-mouse, a recently recognized species found only in Washington and British Columbia.

ASSOCIATIONS WITH STAND AGE

Forest managers (arbitrarily) define successional stages that describe the natural sequence of forest regeneration following fire or logging. Using Appendix 8 in E. Reade Brown's compendium (1985) I summarized the associations of mammals in each of six seral stages in each forest type in western Oregon and Washington and indicated whether the combination of seral stage and forest type offered primary habitat (that habitat upon which a species depends for long-term population viability) or secondary habitat (used by the species but not necessary for its survival).

Most species (Figure 4.2) find primary breeding habitat in either grass-forb and shrub states or large sawtimber and old-growth; the sapling-pole stage supports fewest species. The grass-forb stage provides primary feeding habitat for most mammalian species. Numbers of species decline from thirty-one in the shrub stage to eighteen in old-growth; the fewest species (thirteen) feed primarily in the closed sapling-pole stage.

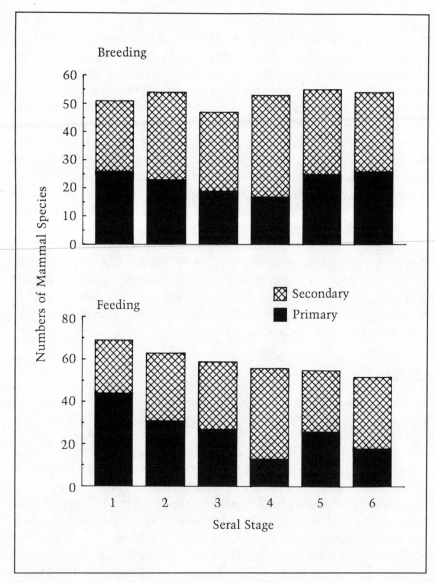

Figure 4–2.   Numbers of mammal species using each of six successional stages as primary or secondary breeding and feeding habitat in western Oregon and Washington. *(Successional stages: 1 = grass-forb; 2 = shrub; 3 = open sapling-pole; 4 = closed sapling-pole-sawtimber; 5 = large sawtimber; 6 = old-growth.) Young closed forest, the stage that dominates managed forest landscapes, is primary breeding and feeding habitat for the fewest species (after Brown 1985).*

Field studies of mammals and their habitat associations show similar trends. I sampled fifty-five species in 166 stands representing six seral stages of Douglas-fir forest in northwestern California. Mean numbers of species sampled per stand were greatest in mature and old-growth stages and lowest in the pole-sawtimber stage. As in Brown's tabulation, diversity was greater in shrubby clearcuts and was similar to diversity in mature and old-growth stands.

Species richness is but one measure of the importance of seral stages for mammals. Abundance is another. Raphael and Barrett (1984) noted progressive increases in abundance of mammals among three late successional stages, with old-growth stands having four times the density of sawtimber stands. We also compared these results with densities in three early successional stages: nine-year-old clearcuts, fifteen-year-old clearcuts, and pole stage timber. Densities in the two earliest stages slightly exceeded those in the old-growth stands, but pole stages appeared to support the lowest densities. Thus, small mammal density follows a more or less U-shaped curve along successional gradients, much like curves for bird diversity, herb and shrub biomass, and quantity of coarse woody debris.

ANCIENT FOREST SPECIES

Research in the Pacific Northwest, although not yet conclusive, suggests that several mammalian species are strongly tied to ancient forest habitats.

The red tree vole is highly specialized, nesting high in the canopy of mature Douglas-fir trees and feeding almost exclusively on Douglas-fir needles. Although tree voles are occasionally observed or trapped on the forest floor, they spend nearly all of their time high in the canopy. In fact, biologists believe that generation after generation might live, reproduce, and die within the same tree.

The western red-backed vole is another distinctive species. Found in mature and old-growth Douglas-fir forests of northwestern California and Oregon, this vole is the only North American mammal that feeds primarily on fungi. It lives in underground burrows, usually near large numbers of well-decayed logs that provide overhead cover and protect the burrows from snow in winter (red-backed voles do not hibernate). Red-backed voles feed almost entirely on truffles, the fruiting bodies (sporocarps) of belowground fungi. Many of these fungi form mycorrhizal associations with roots of conifers, thus enabling the conifers to assimilate soil nutrients and to grow more efficiently. Chris Maser and his associates (1978) have shown that spores excreted by voles that have fed on sporocarps are viable, and he speculates that red-backed voles may be important agents in the dispersal of mycorrhizal fungi.

Red-backed voles can survive on clearcuts unless most slash is removed. Even with plentiful slash and brush, however, red-backed voles are only 50–25 percent as abundant following clearcutting as in ancient forest. This has caused speculation that logging, by reducing numbers of fungus-eating voles, may lead to poorer dispersal of mycorrhizal spores and, ultimately, poorer tree growth.

The northern flying squirrel is found throughout the boreal forest zone of North America. It is unique, however, because it is the only small mammal that feeds upon epiphytic lichens in winter. Like red-backed voles, flying squirrels also dig out and consume hypogeous fungi (spring, summer, and autumn). Studies in Alaska report a strong association of flying squirrels with ancient forest stands: Squirrels feed and reproduce more successfully in the oldest stands. Flying squirrels are less dependent on ancient stands in Washington, Oregon, and northern California, where they are absent from cutover stands but are equally abundant in all forested stages.

The dusky-footed woodrat occurs in western Oregon south through western California. In northwestern California it reaches peak abundance in older, brushy clearcuts where it builds large, woody nests among clusters of regenerating tanoak and Pacific madrone sprouts. The dense cover in these clearcuts protects woodrats from aerial predators. Woodrats are also abundant in mature and old-growth Douglas-fir stands, especially where large logs are abundant and where tanoak forms a dense understory. In these stands, woodrats build nests under large logs and feed upon tanoak acorns. In contrast, woodrats are scarce in young forest. The dusky-footed woodrat is the primary prey of the spotted owl in California, one reason that spotted owls are so dependent on ancient forests.

Douglas' squirrel (or chickaree), the most numerous squirrel in Douglas-fir forests, is absent from clearcuts and reaches peak abundance in ancient forests. Douglas' squirrels feed on a variety of foods including truffles, but in fall (when seed-bearing cones of conifers ripen), they cut cones, peel away the cone scales, and consume the exposed seeds, leaving a huge pile of cast-off scales at the base of their feeding trees. They may prefer ancient forest because older trees produce greater numbers of suitable cones than younger ones. My studies in California also indicate associations with moister sites, higher elevations, and tanoak understories.

Fishers and martens, members of the weasel family, use all age classes of Douglas-fir but are most abundant in mature and old-growth stands. These carnivores tend to avoid large openings, perhaps to avoid predation by lynx, bobcats, eagles, or great horned owls. Their preference for mature and old-growth forests may be explained by the greater amount of downed wood in older stands. Downed wood provides critical habitat for many of the small mammals that are preyed upon by fishers and martens; it also forms protected resting sites used by both species during winter. These subnivean sites, because they are thermally insulated by wood and snow, may be especially critical on the coldest days during storms.

Habitat associations of bats have received little attention, but recent studies by Don Thomas (1988) in the Washington Cascades and the Oregon Coast Range indicate that several species of bats use old-growth stands more heavily than younger stands. Older stands might be especially critical in the Coast Range, where more reproductive females occur. Thomas's surveys revealed three to ten times more bat detections (a detection is the recording of a bat's high-frequency echolocation call) in old-growth than in younger stands. Much remains to be learned about why bats may be more abundant in ancient forests, but preliminary information suggests the difference may be due to greater availability of daytime roosts—particularly large snags—rather than differences in insect prey. □

REPTILES AND AMPHIBIANS

From the time of their remarkable feat of colonizing the land in the Mid-Devonian, nearly 350 million years ago, amphibians have ... undergone a remarkable adaptive radiation, and the living groups exhibit a greater diversity of modes of life history than any other group of vertebrates.

(William E. Duellman and Linda Trueb, *Biology of Amphibians* [1986])

The term "cold-blooded" is something of a misnomer; cold-blooded animals (poikilotherms) are not necessarily cold. A lizard in the desert sun can be warmer than a bird or mammal, so warm that it will die unless it can find shade. But the only way a poikilotherm can stay warm is by being in warm air or water or in sunlight strong enough to warm it. That is why turtles and lizards bask in the sunlight.

Body temperature can have a lot to do with the success of animals. As long as mammals or birds can find, eat, digest, and metabolize enough food, they can stay warm enough to maintain high activity levels even in low temperatures. In contrast, some poikilotherms, including reptiles, are active only if they are warm. At low activity levels they have trouble capturing warm-blooded (homiothermic) prey, avoiding homiothermic predators, and gathering food as effectively as their homiothermic competitors. In the chill of morning, the early bird might catch the worm while the early lizard is torpid.

The Westside has fewer reptile species than the Eastside or areas farther south. Winters do not limit reptiles: Much of the inland West has colder winters yet more reptiles. Rather, low summer temperatures are limiting. Western Washington is seldom warm enough, so reptile species are few and increase only a little through western Oregon to northwestern California. For the same reason, reptiles are scarce in ancient forests. Sunny clearcuts are better places to find them.

Amphibians are quite another matter. Amphibians are also poikilotherms, but they require more moisture and, in many cases, cooler temperatures. In the United States, amphibians are most diverse in the southern Appalachians, where summer temperatures are moderate and rainfall is high. The West, with its hot, dry summers, is not great amphibian country, but there are more species west of the Cascades than anywhere west of Texas. The seeps, streams, and big, moist, downed logs of Westside ancient forests provide excellent habitats for amphibians that cannot tolerate dryness.

And the amphibians of the Pacific Northwest are remarkable. Tailed frogs and three species from New Zealand are the only members of the Leiopelmatidae, the world's most primitive frogs. This family is extraordinarily ancient, dating from Jurassic times, 160–180 million years ago. Tailed frogs are the last of a group once more widespread and are the only North American frogs that fertilize their eggs internally. Their tadpoles have mouths specialized as sucking disks, allowing them to cling to rocks in chilly torrents. And while most toads and frogs take only weeks to metamorphose from tadpole to froglet, tailed frogs need several years.

The Northwest also boasts a family of salamanders found nowhere else in the world. The three species in the Dicamptodontidae occur only

Paul Stephen Corn

Olympic salamander, Olympic National Forest (Washington). *Most sala-manders need cool, moist conditions. This species is endemic to springs and streams in Westside forests and is most common in ancient forests. The larvae of Olympic salamanders are vulnerable to sedimentation; populations take decades to recover after logging.*

west of the Cascades or in northern Idaho. Pacific giant salamanders are among the most important predators in small streams. Michael Murphy and James Hall (1981) of Oregon State University found that they consti-tute 99 percent of the predator biomass in some small streams. Their rare relatives, Cope's giant salamanders, are among the handful of sala-manders that seldom metamorphose to adulthood, reaching sexual ma-turity and breeding as larvae. Olympic salamanders are also dwellers of small streams and require several years to reach maturity.

Some Westside amphibians, such as Pacific treefrogs, are widespread, but a number live only west of the Cascades, and some, such as Larch Mountain and Siskiyou salamanders, are narrowly distributed en-demics (species that occupy extremely small ranges).

Although they are usually in hiding, amphibians are very important in their ecosystems. They can be the most abundant land vertebrates in ancient forests and are probably major predators of small insects, spi-ders, and slugs on the forest floor, especially in rotting logs.

No northwestern amphibian is confined to ancient forests, but several seem to need habitat elements that are most prevalent in old-growth. In the southern Washington and central Oregon Cascades, Bruce Bury and Steve Corn (1988) of the U.S. Fish and Wildlife Service sampled young (less than ten-year-old) clearcuts and three age classes of natural stands: young (30–76 years old), mature (105–150), and old-growth (195–450). They also sampled along a moisture gradient among the old-growth stands. Ensatina salamanders are less abundant in mature forests than in either young or old-growth forests and are most common where there are old, decayed logs and large trees. Oregon slender salamanders seem most common in decayed coarse woody debris. Patterns are weaker for other amphibians except for Pacific treefrogs, which are clearly most abundant in clearcuts.

The picture is roughly similar in northwest California. In Klamath, Six Rivers, and Shasta-Trinity national forests, Marty Raphael and Reginald Barrett (1984) found no trends among total amphibian species in mature and old-growth Douglas-fir/tanoak stands. The only sala-mander showing a clear trend is ensatina, which is only 3 percent and 43 percent as abundant in 119- and 205-year-old stands as in 277-year-old stands. As in Washington and Oregon, California ensatinas are less abundant in mature than in old-growth stands.

A more recent paper by Raphael (1988) examines relative abundances in the same area across all successional stages. Among species for which sufficient data were available, Del Norte and black salamanders reach peak abundance in old-growth, whereas none of the frogs or reptiles does.

Why are differences among natural stands of different ages relatively small? Bury and Corn point out that many "old-growth" features also occur in the young and mature natural stands they studied, including complex structure, snags, and lots of downed logs, particularly in the older decay classes. This is how they explain these similarities:

Wildfire often burns unevenly through stands, resulting in patches of lightly burned or unburned vegetation surrounded by areas more intensively affected by fire. Some large trees might not be killed during fires and these persist into the regenerated stand. Burned trees become snags that later fall to the forest floor, creating huge amounts of CWD [coarse woody debris]. This heterogeneity and large amounts of CWD in naturally regenerated forest likely maintain favorable conditions for many species of the herpetofauna.

Managed stands (clearcuts) had little downed CWD in older decay classes ... and, generally, no snags nor trees (except for a rare spar pole or small planted trees). Current forestry practices usually fell all trees and snags at sites, eliminating variability in stand age and structure. Logging is generally followed by prescribed burning of slash and cull logs, reducing CWD by 50% or more. (1988, p. 19)

Thus, differences among successional stages of natural forests are small compared with differences between natural forests and tree plantations. Management practices that eliminate downed logs are not good for terrestrial amphibians. And as Corn and Bury (1989) have found (see Chapter 6), logging practices that increase sedimentation in low-gradient headwater streams are no less damaging for aquatic amphibians.

SALMONID FISHES

The ancient forest ecosystems that produced the trees and soils undergirding two major Westside industries, timber and agriculture, support yet another. Salmon was the resource around which the first Northwesterners built their cultures. Nutritious, easily caught, and phenomenally abundant, their dependable migrations determined where Native Americans lived and what they did from week to week. And when Euro-Americans settled, they found the waters draining the ancient forests from Alaska to northern California to be the world's richest salmon factories.

Although less abundant now, native members in the family Salmonidae still provide the region's premier fisheries. Chief among them are five species of Pacific salmon: chinook or king, coho or silver, chum or dog, red or sockeye, and pink or humpy; and three species of trout: rainbow, cutthroat, and Dolly Varden char.

Some Westside salmonids reside in lakes or streams. But with the exception of landlocked sockeyes, the salmon are anadromous: Although they are born in freshwaters, they do most of their feeding and growing in the sea, as do some populations of trout. Migratory salmonids are vulnerable throughout their lives. As sexual maturity approaches, fish that slip past high-seas fishermen and orcas migrate to river mouths and, ultimately, to the same streams where they were born. The ones that evade riverine fishermen and get around dams then compete for mates, lay their eggs, and die. Those eggs that escape being eaten or smothered hatch into young that remain for a while in fresh water. And the young that survive in their rearing habitats finally migrate back to the sea, completing the life cycle.

The details vary among the salmon species. Some stay at sea for five years or more, others for less than two. After hatching, the young of some remain in freshwaters for a year, others for only a few days. But they and anadromous trout share two basic requirements away from their marine feeding grounds: a pathway to their breeding streams and conditions suitable for spawning and rearing once they get there.

A price of developing the region's rich hydropower potential has been dramatic loss of salmon production because dams have barred salmon from reaching their spawning streams. In some cases, dam builders installed "fish ladders" to provide a pathway, although ladders are not always effective. Artificial spawning and rearing compensate for a small part of the loss, but the facilities, energy, feed, and labor are not free, unlike the natural services they replace. People—usually taxpayers—pay for them.

Even where salmon can reach mating grounds, they need the right kind of substrates, plentiful, cool, well-oxygenated fresh water in which to spawn and undergo early development, and food (mainly stream insects). Such conditions were abundant when the Westside landscape was covered with ancient forests. They are far less abundant today. A major reason is a drastic reduction of logs in streams.

In an ancient forest, small and medium-sized streams are crossed by many large downed logs. Larger streams have accumulations of logs along banks, often at bends. As beautifully explained by Fred Everest and coauthors (1985) and by Forest Service researchers Jim Sedell and Fred Swanson (1984), logs affect ancient forest streams in many ways. One of the most important is by serving as baffles that modify stream-flow, channeling, speeding, and (especially) slowing it. This creates varied habitats, from plunge-pools and fast-flowing riffles to deeper, quieter pools and side channels. Logs create the characteristic "stair-step" appearance of smaller streams. They are less obvious in larger streams but are very important there as well.

Logs increase structural complexity in streams, providing cover and affecting flow regimes. Different salmonid species and life history stages need currents of different speeds. Current speed affects dissolved oxygen concentrations, temperatures, the kinds and numbers of food organisms, and the amount of energy that salmonids must spend to maintain their position. The diverse current regimes in ancient forest streams allow salmonids—from delicate fry to powerful adults—to select habitats appropriate to their particular needs. Streams in younger natural forests often have abundant large logs as well. One reason is that

Tim Crosby

Logs in creek, Mt. Baker-Snoqualmie National Forest (Washington). *The large logs from ancient forests form baffles and dams that create pools, gravel beds, and other stream habitats. In their absence, streams have less habitat diversity and fewer salmon and trout species.*

logs usually survive fires that initiate young stands. Another is that large waterlogged logs decompose slowly. Thus, these logs can endure until new ones begin entering streams.

Structural complexity increases species diversity. More kinds and life history stages of salmonids live in ancient forest streams than in streams where large logs are lacking.

The amounts of logs in streams have declined dramatically because logs were seen as hazards to navigation and salmon migration, because they were flushed downstream by torrents released from loggers' splash dams, and because forest management has decreased recruitment of large logs to streams, since trees are cut before they die and fall over.

Streams through logged areas often have few or no large logs. Consequently, they have few pools and consist mainly of riffles. With currents unobstructed, flow rates are often high. As Peter Bisson of Weyerhaeuser and Jim Sedell (1984) have shown, in summer, such structurally simple habitats support mainly salmonids—especially underyearling steelhead (anadromous rainbows)—that do well in the fast riffles. They prosper, but older steelhead, cutthroat, and coho do not.

Logs can be especially important in winter, when flow rates are highest. Major storms can sweep salmonids downstream to their deaths. Streams surrounded by clearcuts or tree plantations have little protection from scouring by storm-generated currents. But logs and root wads in ancient forest streams create quieter refuges where salmon and trout can congregate, allowing them to survive storms.

Logs affect the particle sizes of sediments. Flowing water can move rocks as large as peas, limes, or even watermelons, but it deposits them wherever it slows too much to keep them suspended. Not surprisingly, the coarsest, heaviest particles drop out first. By slowing the flow, logs tend to accumulate gravel nearby, forming ideal salmon spawning beds.

Salmon lay eggs in depressions (redds) they have dug in streambed gravels, then cover the eggs by fanning more gravel over them. Their choice of well-sorted gravel is no accident. The particles are small enough for salmon to move, but the pore spaces among them are large enough to allow streamwater to circulate around the eggs. Circulation is essential because developing salmon embryos need a high oxygen concentration. Circulation decreases when finer sediments—sands and silts—clog the interstices among gravel particles, slowing growth of the developing embryos and increasing mortality.

Little sand and silt enter ancient forest streams because roots and stems prevent their movement. But timber operations—particularly road building—dramatically increase the amount of fine sediments entering streams, by ten, 100, even 1,000 times. This is not only true where entire watersheds are cut; in a western Oregon watershed where 70 percent of the land was uncut and buffer strips of trees were left along stream margins, stream sediment loading still increased enough to degrade spawning beds. Increased suspended sediments also cloud the water, inhibiting foraging by salmonids, which are visual feeders. At high levels, sediments damage salmonid gill tissues.

Timber operations not only increase suspension and deposition of fine sediments; they can also eliminate gravels needed for salmonid spawning. Forest cover promotes infiltration rather than runoff, and the abundance of logs in ancient forests slows currents in streams. Together, these tend to smooth out peak flows in streams. But in heavily roaded and logged watersheds, major winter storms can sluice all sediments (including gravels) from streambeds, leaving nothing but bedrock.

Ancient forests also increase minimum summer streamflow compared with young forest, as explained by Richard Myren and Robert Ellis (1984) of the National Marine Fisheries Service. Rapidly growing young forest transpires more than old-growth. In summer periods of low flow, the increased flow in ancient forest watersheds could be crucial.

Ancient forests also affect stream temperatures. Westside salmon and trout prefer temperatures from 5 to 15° C (41–59° F), but are stressed at higher temperatures and die when they reach 23–26° C (73–79° F). Because salmonids need high concentrations of oxygen, high temperatures put them in a double bind. They both raise the fishes' metabolic rates and lower the amount of dissolved oxygen in the water and can thus cause suffocation. Salmonids are also more vulnerable to epidemics in warm waters; indeed, diseases can kill them at temperatures well below the point of direct lethality.

Shade prevents streams from overheating during summer when the flow is low. As Dick Fredriksen and Dennis Harr (1981) note, the size of this effect can be large: after logging and slash burning, a stream in the Oregon Coast Range reached a maximum of 29° C (84° F), 15° C above its predicted maximum under forest cover. Whereas small trees can shade the smallest streams even when the sun is high, such streams usually have few salmon and trout. The larger streams that can have large populations of diverse salmonids, however, can be shaded only by larger trees. As Robert Beschta and coauthors (1987) point out, the importance of high temperatures on salmonids is complex. Clearly, it is less a problem on the Olympic Peninsula than in southwest Oregon and northwest California, where temperatures can approach and exceed lethal temperatures unless streams are well shaded.

Beside keeping streams cool, conifers also provide most of the food in smaller ancient forest streams via litterfall. Conifer needles and twigs are slow to decompose and provide food for insects that salmon eat, but the large logs across streams retain substantial amounts of this detritus until it can be eaten. Deciduous red alders and willows in the riparian zone also provide abundant, fast-decaying litter. And when large conifers die, more sunlight reaches the stream, increasing production of algae that are grazed by aquatic insects.

Logging damages salmon populations in many ways but can have one beneficial effect (at least in cooler areas). Eliminating the canopy over a stream increases light levels, thereby raising algal production more than enough to compensate for the loss of litterfall from the trees. However, as Sedell and Swanson (1984) explain, this effect generally lasts less than twenty years (Figure 4.3). After canopy closure in a tree plantation, there is less food for salmonids throughout the rest of the rotation—that is, most of it—than there was in ancient forests. This echoes the effect of current logging practices on forage for deer, where an improved short-term situation is obtained at the price of long-term worsening.

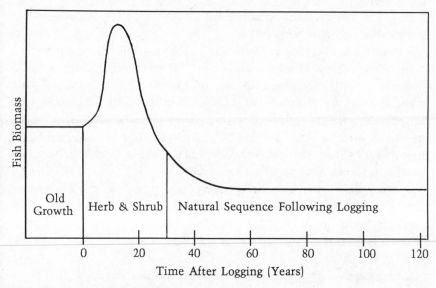

Figure 4–3.   Effects of logging on salmon and trout biomass. *Biomass increases briefly after ancient forests are logged because more light reaches the stream, encouraging the growth of algae, which are eaten by insects, which, in turn, are eaten by fishes. But after the canopy closes—for most of the rotation—the biomass of these fishes drops below levels found in natural forests (from Sedell and Swanson, 1984, p. 12).*

Logging and dams have profoundly altered Westside streams in this century. It is possible that salmonid production can be increased by intelligent manipulation of their habitat in conjunction with timber operations. But recent findings suggest the model that managers should emulate to improve salmon runs. It is the streams produced by ancient forests.

PLANTS

In 1977, the annual U.S. retail value of cascara sagrada reached $75 million. . . . The young men we interviewed in Washington said on a good day each of them could bring in 280–300 pounds.
(Christine and Robert Prescott-Allen, *The First Resource: Wild Species in the North American Economy* [1986])

Giant conifers loom large on our mental landscapes. Few living things have such ability to inspire and humble human beings. But in the shade below ancient forest canopies are plants less imposing, to be sure (see the essay below by Jean Siddall and Rick Brown), but possessing some remarkable qualities nonetheless.

Besides their inherent value, some of these species have significant commodity values. One, a small broadleaf tree called cascara buckthorn, has a rather unglamorous use. Westside Indians knew it well, and in the last century Northwesterners have stripped its bark to provide the pharmaceutical industry with cascara sagrada, which has been called the world's most widely used cathartic laxative. Cascara buckthorn is very shade-tolerant, and scattered individuals occur in both old-growth and some second-growth forests.

Another plant with potentially much greater medical value is a small tree or large bush with dark green foliage like that of a redwood or hemlock. Some are quite handsome, but the word that comes to mind for most is "scruffy." The charms of Pacific yews emerge mainly as you get to know them.

But they are worth knowing. Many ancient forest songbirds find them worthwhile, for yew seeds are not borne in cones and dispersed by the winds, as are those of other Westside gymnosperms. They are enfolded in fleshy, berrylike arils, whose bright red color attracts birds that eat them and disperse the hard seeds throughout the forest. Roosevelt elk and Columbian black-tailed deer also find yew foliage worthwhile; many yews get their scruffiness from being heavily browsed.

Although their looks might not command attention, Pacific yews have inner beauty; their heartwood is bright orange or rose-red, dense, hard, very fine-grained, and strong, yet amazingly flexible. These qualities make it wonderful for fine cabinetry and made English yew the wood of choice for European bow makers. Robin Hood and his merry men knew well the virtues of yew trees.

For people today, the most remarkable thing about Pacific yews is their bark. It is unusual to the eye, true enough: dark red or purplish, occurring in very thin, long, wavy strips. But it is the yew bark's chemical composition that has scientists so excited, for it contains a compound—taxol—that could help humankind to fight one of our most feared enemies: cancer.

Cancers (more than 100 different kinds in humans) are diseases of uncontrolled cell division. Taxol blocks cell division by a mechanism unlike that of any compound studied to date. As Matthew Suffness (1985) of the National Cancer Institute states, taxol "is a very valuable tool . . . and has been used quite extensively in the past two or three years. . . . Taxol has a fairly broad antitumor spectrum and because of its novel mechanism is a drug which is of high interest for clinical study." Early studies showed that taxol has promise for treating leukemia, colon, lung, mammary, prostate, and pancreatic cancers. Clinical trials began in 1984.

But before taxol's anticancer potential can be explored fully, the National Cancer Institute needs ample supplies. Taxol occurs in minute quantities in Pacific yew bark; 30 tons of bark stripped from 12,000 yew trees yield only 5–6 pounds of taxol. And taxol has defied attempts at synthesis in the laboratory for more than a decade. Pacific yews will be the only source of clinically useful amounts for years to come.

Unfortunately, yews have rather finicky habitat requirements and are much less common than they once were. These very shade-tolerant trees grow exceptionally slowly, which limits them largely to ancient forest, as the Forest Service's Tom Spies has recently shown. So much ancient forest is now gone that it will be difficult to get enough bark for clinical trials, let alone enough to treat millions of cancer patients. Between losing their habitats and being stripped of their bark, Pacific yews could become the spotted owls of the plant world.

Of course, if forests were managed to maintain biological diversity, there would be many more yews. But loggers have always ignored yews, either burning them or leaving them to rot. And when the Forest Service and the BLM attempt to restock cutover lands, they plant fast-growing species valued for timber, not those that might cure cancer. Yews grow so slowly that planting them now would not soon pay dividends anyway; yews 100–200 years old are often only 6 inches in diameter. Small yews yield little bark.

The only way to have enough yews in years to come is to protect our remaining ancient forests and to manage lands where logging is allowed with far more care than they are now given. We can do this only if Congress, foresters, and citizens recognize that forest management is more than timber management. Managing for biological diversity is essential to maintain our options for the future. We will never know how many potential cures for cancer, heart disease, or AIDS have already been clearcut to extinction.

Perhaps the shortage of Pacific yews will not affect us. Only 30 percent of us will develop cancer. Perhaps you and I, our mates, our children, our parents, and our friends will all be lucky.

Perhaps not. But even if there isn't enough taxol to go around, perhaps scientists will discover an equally potent anticancer drug in some plant from the ancient forests of Brazil. That would be possible, if only Brazilians were not eliminating their ancient forests even faster than we are eliminating ours. Then again, it is possible that the Northwestern senators and congressmembers whose decisions are eliminating our ancient forests can persuade Brazil to save its ancient forests for our benefit. Possible, but not very likely.

Plants of Westside Ancient Forests

☐ *Jean L. Siddall, Native Plant Society of Oregon, and Oregon Rare and Endangered Plant Project, Lake Oswego, Oreg., and Richard T. Brown, National Wildlife Federation, Portland, Oreg.*

Are there species of plants that grow *only* in ancient forests? Probably not, but two groups are known to rely heavily on old-growth. The first includes species found primarily in ancient forests. The second includes other species that rely on old-growth components such as large snags and downed logs, which are produced only in ancient forests but persist long after an old-growth stand burns or blows down. These components provide essential habitat as the forest regenerates.

For the first group, ancient forest ecosystems provide the needed long-term stability and relatively constant levels of shade and moisture for slow-growing species such as Pacific yew. Other members of this community include shade-tolerant herbs that spread slowly by rhizomes or other vegetative means. Notable examples include Mt. Hood bugbane and Hall's isopyrum, which flourish in moist areas under old-growth Douglas-firs but appear very stressed in clearcuts. Unfortunately, scientists do not know whether they can survive once the canopy has been removed.

Other slow-growing, rhizomatous species such as wintergreens, rattlesnake orchid, prince's-pines, and coralroots are also typical members of ancient forest communities. Although they also occur in other forest types, the most abundant and vigorous populations are found in ancient forests.

At least one nonvascular plant is characteristic of virgin coniferous forests of the Pacific Northwest: *Scapania bolanderi*, presently a common liverwort apparently known only from the bark of old trees. It does not grow on young trees and will disappear if the old trees are logged. The second plant group includes species that rely on old-growth components—large downed logs and snags—that may be carried over into regenerating forests after a fire or windstorm. These ancient forest "legacies" are essential to the continuance of this plant community.

The best-known example is the western hemlock seedlings that grow on "nurse logs." Other species that occur primarily, if not always, on rotting wood of previous old-growth trees include spreading wood fern, wood nymph, and kruhsea. Kruhsea and wood nymph also appear to develop their most vigorous populations within existing ancient forest stands.

As old-growth logs decay further and become incorporated into the soil, they support a group of nongreen vascular plants. These members of the heath and orchid families were formerly referred to as "saprophytes," suggesting that they live by breaking down dead organic matter. Increasingly, they are called "mycotrophs" (from the Greek *mycos* [fungus] and *trophos* [nutrition]) because they derive their sustenance from fungi. There is evidence that the fungal hosts of at least some mycotrophs occur only in large decomposing logs that have been incorporated into the soil. Only old-growth forests produce these essential substrates.

In summary, although no plant species are currently known to occur only in old-growth, there certainly are species that reach peak abundance in ancient forests. Much more research is needed to determine how fragmentation affects such stands, not only the plants, but also their pollinators, herbivores, and other members of the ancient forest community. Especially in view of such threats as global warming, ancient forests may provide the only remaining stable habitat conditions necessary for these species to persist. □

FUNGI

. . . how do we deal with the values of organisms whose very existence escapes our notice? Before we fully appreciated the vital role that mycorrhizal symbiosis plays in the lives of many plants, what kind of value would we have assigned to the tiny, threadlike fungi in the soil that make those relationships possible?
(David Ehrenfeld, "Why Put a Value on Biodiversity?" in Edward O. Wilson, ed., *Biodiversity* [1988])

If there is an earthly paradise for fungi, I have seen it, felt it, drawn its rich perfume into my lungs. From what I have seen, it is not tropical rainforests. The tropics are mycologically so little studied that they could well prove to be far richer; nobody knows for certain. But fungi are less conspicuous there. In moist Westside ancient forests, fungi are everywhere.

Their ubiquitousness might dismay some people, those who associate fungi with disease, sickness, and death. But this view misses the infinitely more complex truth. Fungi are amazing organisms.

In most cases people overlook them. When we think of living things, we usually mean plants and animals and either include fungi with the plants or ignore them. But fungi have their own kingdom, as different from vine maples as from cougars. Ignoring them overlooks some of the most vital components of terrestrial ecosystems, especially ancient forests. One must work hard to overlook fungi in ancient forests.

Two people who do not are Bill Denison and Jim Trappe, who share their special insights, born of decades of research, into the fascinating, mostly hidden world of fungi in the following essays.

Fungi of Ancient Forests: Hundreds of Mushrooms, Legions of Molds

□ William C. Denison, Department of Botany, Oregon State University, Corvallis, Oreg., and Northwest Mycological Consultants, Corvallis, Oreg.

Enter an old-growth forest a few weeks after the rains have started in the autumn. Mushrooms have thrust up through the forest floor and popped out from the tops and sides of fallen logs. Mushrooms in a bewildering variety of colors, shapes, sizes, and textures seem to be everywhere. There are red and yellow ones, brown ones and white; some are even violet, green, blue, or black. Some have the typical umbrella shape. Others look like corals, blobs of Jell-O, or hooves.

Each mushroom is shedding spores. You might see a dusty deposit of white or cinnamon where the spores have fallen to the ground, but most of them have been borne away on the breeze to start new colonies of mushrooms. The purpose of a mushroom is to produce spores, which, like seeds, multiply and spread the fungus. In a few days, most of the mushrooms will decay or be eaten, but the fungi from which they came will live on.

The greater part of each fungus is invisible. The main part, its *mycelium*, consists of a network of fine threads (*hyphae*). The hyphae grow within the soil or log, "feeding" by secreting digestive enzymes that break down nearby tissues of dead plants so they can be absorbed. By feeding, fungi decay organic matter. You have undoubtedly seen the cottony mycelium of fungi decaying food you had intended to eat yourself; mold is one kind of mycelium.

The mycelia of some fungi produce mushrooms just as an apple tree produces apples. But the mushrooms you see give only the slightest clue to the hordes of fungi living there. Pick up a handful of decaying litter from the forest floor, look closely, and you will find fungal threads knitting some of the fragments together. You might see minute fruiting bodies of fungi that do not produce mushrooms. You will certainly see twigs and leaves in various stages of decay, decay caused mostly by fungi. Many of the tiniest animals, from mites to amoebae, feed on decay fungi just as cows graze on grass.

Now pull away the bark of a fallen log. If the log has been down for decades or more, the wood will probably be so soft that you can pull it apart with your fingers. This too is the work of fungi, which have been feeding on the solid material of the wood, thereby decaying it. In the log are insects and other minute animals that feed primarily on decay fungi.

These almost invisible fungi have the essential role of recycling the products of the forest. As the wood and leaves are broken down, their ingredients are made available to the next generation of plants and animals. But if the most important role of fungi in forests is decay and recycling, it is far from their only role. Some form symbiotic associations, called *mycorrhizae*, with the roots of forest trees. Most timber trees require such a fungal partnership in order to grow normally. Other fungi are lichens, living in the tops of trees, providing nitrogen to the forest. Still others, the *endophylls*, live inside normal healthy leaves, where they produce chemicals that make the foliage distasteful or poisonous to insects. A few fungi attack trees and other forest plants, causing diseases.

In the paragraphs that follow, I explore the diverse roles of fungi in ancient forests by describing the lifestyles of six species: the chanterelle, the Oregon conk, the Douglas-fir endophyll, the lichen *Lobaria oregana*, the laminated root-rot fungus, and the Oregon white truffle.

THE CHANTERELLE, A GOLDEN HARVEST

Every autumn, after the rains begin in the Pacific Northwest, a golden-yellow mushroom, the chanterelle, appears in the forest. Chanterelles are eagerly sought as gourmet edible mushrooms. In recent years, millions of dollars worth of chanterelles have been shipped to the cities of North America and Europe to grace the tables of the affluent.

In the Westside, many people, including laid-off mill hands and loggers, supplement their unemployment checks by picking and selling wild chanterelles. In a good year up to 7 million pounds of chanterelles are picked in Oregon and Washington and sold out-of-state, producing millions of dollars of income, much of it going directly to those small, rural communities most affected by the decline of the timber industry.

Chanterelle mushrooms are the "fruit" of the chanterelle fungus, whose mycelium lives in forest soil and forms mycorrhizae with the roots of Douglas-firs and other trees. A mycorrhiza is a special root formed jointly by plant and fungus. Mycorrhizal roots look markedly different from nonmycorrhizal roots of the same tree. Nonmycorrhizal roots are dark, opaque, and consist of a central tap root with side branches. Mycorrhizal roots are pale, almost translucent, and fork repeatedly into two equal branches. The exterior of a mycorrhizal root is enveloped in a mantle of mycelium, whose hyphae invade the outer tissues of the root, growing in the spaces among the root's cells.

A healthy tree has both mycorrhizal and nonmycorrhizal roots. When Douglas-fir seedlings are grown in sterilized soil, a mycorrhizal fungus must be introduced into the soil in order to obtain normal seedling growth.

The fungal mycelium extends from the mycorrhiza into the adjacent soil. It undoubtedly feeds to some extent by decaying organic matter, but much, if not most, of its food is provided by its tree partner. In return, the fungus accumulates mineral nutrients (especially phosphorus) from the soil and shares them with the tree. Seedlings grown in the absence of mycorrhizal fungi often show symptoms of phosphorus deficiency, even when the soil has ample phosphorus.

The mycelium of the chanterelle fungus produces an annual crop of mushrooms just as its tree partner produces crops of cones. Concern has been expressed that the commercial harvest of chanterelles will exhaust the supply. The little evidence available indicates that this is not the case. It would probably be as difficult to exhaust the supply of chanterelles by picking the mushrooms as it would be to exhaust the supply of trees by harvesting their cones.

However, clearcutting the forest is a different matter altogether. When a stand of trees is clearcut, the fungus may produce one last crop of mushrooms the following fall. Then it is all over. There will be no more chanterelles until a new forest is well established on that site, typically many years later.

The dependence of the mushroom on the forest is absolute. Without forest trees there are no mycorrhizal mushrooms. As air pollution has damaged the forests of Europe, their chanterelle harvest has declined. To some extent our new chanterelle industry is the result of market demand brought about by the death of European forests.

THE OREGON CONK

Not all fungi live in the soil. The Oregon conk lives in logs and snags. In the language of foresters and loggers, a "conk" is the hard, woody, hooflike fruiting body of a wood-decay fungus. A conk, like a mushroom, is the visible expression of a fungus at work. Like mushrooms, conks produce spores that spread the fungus to new localities. Mushrooms are soft, some are edible, and most decay quickly. Conks, on the other hand, are woody, inedible, and often persist for months or years.

The Oregon conk is a king among conks. It produces some of the largest fungus-fruiting bodies known—sometimes more than 2 feet across and weighing more than 20 pounds. The fungus that produces this monster lives inside the huge logs and standing snags characteristic of old-growth forests. The mycelium attacks and digests the lignin and cellulose of the cell walls of the dead tree's heartwood. Because cell walls give a log its physical strength, the wood softens as they are broken down. Eventually the dead tree topples and the decaying log merges with the soil. Many organisms collaborate in this decay, but the fungi that attack cell walls are essential to the process. It takes a large mycelium to produce a large conk; fungi that decay smaller trees and limbs do not accumulate enough resources. Big conks are a product of big trees.

A large conk normally lasts for years, growing and producing a new layer of spore-bearing tubes on the underside each year. Eventually conks die, are invaded by other fungi, and become homes for fungus-eating springtails, fly and wasp larvae, and beetles.

Every category of wood has certain fungi that specialize in its recycling. Sapwood and heartwood; the wood of conifers, hardwoods, and of shrubs; twigs, branches, limbs, trunk, and roots—each has a special group of fungi that attack its cell walls, converting wood to humus.

THE DOUGLAS-FIR ENDOPHYLL

But not all fungi are engaged in recycling. Our next example, the Douglas-fir endophyll, is an unobtrusive and helpful guest inside healthy foliage.

Most of us are accustomed to thinking of microorganisms such as bacteria and fungi as enemies that, if they enter our bodies, cause disease, or even death. But fungi such as the Douglas-fir endophyll, an inhabitant of healthy Douglas-fir foliage, can be helpful guests in their host's tissues.

Each spring, buds burst and new twigs emerge bearing new needles. Since Douglas-fir is evergreen, its needles remain for several years— commonly seven to eight but occasionally up to eleven or twelve. During this relatively long period, very few insects attempt to eat them. This is remarkable when you consider how much insects can damage the foliage of deciduous trees and garden plants in a single year. Douglas-fir needles are protected in part by a tough outer coating and resins produced by the needle itself and in part by natural insecticides provided by their endophyllic fungi.

A newly emerged needle has no endophyllic fungi, but by the second year endophylls have occupied all or nearly all of the needles. Sometime during that first year, an endophyll spore lands on each needle, germinates, and sends a new hypha growing into its interior. There the fungus develops a tiny mycelium. The mycelium causes no disease, and the needle is—by any standard—healthy and normal. Typically, endophyllic fungi are unable to use six-carbon sugars such as glucose. This appears to be an evolved accommodation that keeps the fungus from interfering with photosynthesis, which is, after all, the primary function of the needle.

The endophyll produces several chemicals that are of no direct benefit to it but are either distasteful or downright poisonous to insects. The presence of these chemicals reinforces the defenses provided by the needle itself, making needles unavailable as a food supply for grazing insects.

The Douglas-fir endophyll does not produce a mushroom or conk. Its fruiting body is minute, scaled down to a size appropriate to a needle and the limited resources of its small mycelium. Furthermore, the fungus does not make fruiting bodies until the needle is senescent or dead, again avoiding conflicts with the needle's primary photosynthetic function.

A NITROGEN-FIXING LICHEN, *LOBARIA OREGANA*

The fungi we have examined so far, though very different, have at least one thing in common: the body of the fungus is a mycelium growing invisibly within the soil or in a root, log, or needle. Our next example is a fungus whose body is a lichen thallus growing out in the open.

Many of the fungi found in an old-growth forest are lichens. Some form a patchwork of colorful crusts on trunk and branches. Some dangle like skeins of pale green hair. Still others form clumps or lettucelike sheets. Each lichen fungus encloses one or more kinds of algae. Their relationship is a classic example of mutualistic symbiosis: two different organisms living together for their mutual benefit. The alga carries on photosynthesis, providing food for both itself and the fungus. The fungus provides certain nutrient elements and protection for the alga.

Because algae need light to photosynthesize, the lichen must grow where it is exposed. Although it is quite possible to pass through the forest without noticing the mycelia of decay fungi, it is difficult to overlook the lichen thalli, the equivalent parts of lichen fungi.

Sometimes lichens are confused with mosses, which are also abundant in ancient forests. But if you look closely, you will see that mosses (or their relatives the leafy liverworts) are made up of minute stems and leaves. Lichen thalli are diverse in form but never consist of stems and leaves. Mosses often carpet the forest floor and form clumps and tuffets on tree trunks, especially where moisture levels remain high for much of the year. The principal domain of the lichens, on the other hand, is in the old-growth canopy. They thrive on twigs and branches high above the ground.

In ancient forests on the west side of the Oregon Cascades, the most abundant lichen is *Lobaria oregana*. Growing in treetops and often hidden by lower branches, it may be nearly invisible from the ground. Nevertheless, it is usually present in quantities approaching a quarter ton (dry weight) per acre. Typically, its weight is greater than the combined weights of all the other lichens, mosses, and liverworts growing

Elliott A. Norse

Lobaria oregana, **an ancient forest lichen, on a nurse log, Olympic National Park (Washington).** *This slow-growing species occurs in the canopies of Pacific Northwest forests that are more than 100 to 150 years old. Blue-green bacteria within the fungus convert atmospheric nitrogen into compounds that plants need to make proteins. Carried to the forest floor, these nitrogen compounds are essential to maintaining soil fertility. Intensive forestry eliminates* Lobaria *and the other major sources of nitrogen.*

on the trees. In fact, its biomass in an old-growth Douglas-fir is roughly one-fifth the biomass of foliage on the same tree.

This abundant lichen starts life as a *lobule,* a mere fleck of lichen, weighing about 10 milligrams. A mature *Lobaria* thallus produces thousands of lobules, literally "chips off the old block," each containing both the fungus and the algae needed to establish a new lichen. The lobules break loose from the parent and blow about; if one settles in a suitable spot, it will attach and grow.

Lobaria lobules grow slowly, as do other lichens. In five years a lobule may be as big as your little fingernail: in fifty years it might reach the size of a small head of lettuce.

Lobaria is a foliose (leaflike) lichen. Its upper surface appears green when wet because just below the surface is a layer of green algae. The alga (*Myrmecia*) grows, carries out photosynthesis, and leaks a little sugar (ribitol) to its fungal partner. Beneath the algal layer, the thallus is white and cottony, consisting mostly of fungus. Embedded in this layer, like raisins in a cake, are small, spherical nodules containing *Nostoc*, a cyanobacterium (bluegreen alga).

Lobaria oregana, like some other lichens, is a three-way partnership involving the fungus, the green alga, and the cyanobacterium. *Myrmecia* is the principal photosynthetic partner; *Nostoc* fixes atmospheric nitrogen.

Nitrogen is essential to the growth of all living things. Although it is the most abundant gas in our atmosphere, gaseous nitrogen (N_2) is not directly usable by plants and animals. Before it can be used, it must be "fixed," that is, chemically reduced, then combined with other elements, notably hydrogen (as ammonia) or oxygen (as nitrate). Relatively few living things can fix nitrogen: only a few bacteria and cyanobacteria, such as the *Nostoc* found in *Lobaria*.

All natural ecosystems depend upon the handful of microorganism species that fix their nitrogen. In ancient forests, lichens with cyanobacteria (such as *Lobaria*) are often important sources of fixed nitrogen. In a typical Cascades old-growth coniferous forest, *L. oregana* fixes 5–10 kg of nitrogen per hectare (5–9 lb./a.) per year. The lichen uses some of the fixed nitrogen for its own growth; some leaks to the environment. The nitrogen is eventually released to the rest of the ecosystem when the lichen dies and decays.

A FUNGUS THAT ATTACKS TREES, *PHELLINUS WIERII*

The fungi discussed up to this point are beneficial to the forest in one way or another. Our next fungus, *Phellinus wierii*, is a pathogen, a disease-causing fungus. It kills certain trees. From a timber producer's perspective, at least, it is a villain.

A few fungi attack living plants, including forest trees, causing catastrophic diseases. In very rare instances an introduced fungus disease may threaten to wipe out an entire tree species. The fungi causing chestnut blight (introduced from Asia) and Dutch elm disease (introduced from Europe) are two examples; another alien fungus called *Phytophthora lateralis* now threatens Port Orford cedar. The laminated root rot fungus *Phellinus wierii* is a major pathogen, yet it is a native species that does not threaten to wipe out its victims, which include Douglas-fir, true firs, hemlocks, and spruces. *P. wierii* affects the roots of living trees. When enough roots are destroyed, the tree may die where it stands or crash to the ground.

Oddly enough, *P. wierii* rarely produces fruiting bodies and spores, and when they are formed, they often occur at ground level or just belowground, so that spores have little opportunity to blow away to a new site. Even odder, attempts to infect trees deliberately, using spores of *P. wierii*, have been unsuccessful. Evidently, trees seldom get infected by *P. wierii* spores. If not by spores, then how?

Long after a tree has died, the mycelium of *P. wierii* continues to grow in its roots and stump. The mycelium doesn't grow very far out into the soil, but if the root of a susceptible tree grows alongside an infected root, the healthy root is invaded. Thus, the fungus spreads by direct contact of a healthy root with infected wood.

Initially a single tree is killed. Then other nearby trees that are susceptible also become infected. Gradually, over many years, a patch develops in which most susceptible conifers are dead or dying. Such a patch may cover an acre or more and involve dozens of trees. New trees planted in the patch are quickly infected. Cleaning up the patch by removing all the buried, infected wood would be an impossible task. From the viewpoint of a timber-oriented forester, this disease is a disaster.

Many foresters feel that their job is to grow trees, to grow them right now, to grow them profitably, and to grow them on the particular piece of ground that is his or her special responsibility. However, there are other viewpoints. One might ask whether, in the long run, it is wise policy to convert our public forests to tree farms. Other inhabitants of unmanaged forests may be as valuable as the timber of susceptible trees. Some, like the chanterelle and unsusceptible trees such as western redcedar and incense cedar, have considerable economic value. Others, such as the lichen *Lobaria*, have ecological value: They play subordinate but essential supporting roles. Certainly, we do not yet know enough about any of the plants, animals, or microorganisms of ancient forests to write off any one of them as expendable, even such obvious pests as mosquitoes and *P. wierii*.

Is laminated root rot an unmitigated disaster when viewed in this broader perspective? It is not a new disease. The fungus has existed throughout the forests of the Pacific Northwest since before the first European settlers, probably long before. Somehow forest and fungus managed to coexist. To understand how, we need to look at the course of the disease in an unmanaged forest over a much longer time than commercial forestry allows for the growth of a crop of trees.

When an old tree is felled by *P. wierii*, its remaining roots are usually heaved up out of the ground. Many of them dry out and, since they are aboveground, no longer serve as sources of infection for the roots of adjacent trees. The upheaval caused by this uprooting of the tree loosens and aerates the soil, encouraging fungi that compete with or suppress *P. wierii*. Younger trees killed by the fungus are likely to die standing; their roots remain in the soil to infect other trees. The spread of infection can be broken where it encounters an old tree or an individual of a species that resists laminated root rot. Consequently, *P. wierii* spreads more slowly through old-growth stands than through younger stands, especially young Douglas-fir monocultures.

In an old, unmanaged forest, a laminated root rot infection spreads like a very slow ripple. Over the course of centuries, it breaks up or passes out of the forest. Where the fungus is abundant and active, all susceptible conifers are killed and new seedlings quickly succumb. This creates openings where various shrubs, hardwoods, and cedar seedlings, unaffected by *P. wierii*, become established. In time, the stumps, logs, and large fragments of conifer wood decay completely, and *P. wierii* dies out. Seedlings of susceptible conifers can now reestablish and survive. Meanwhile, *P. wierii* is present elsewhere in the forest. The outer edge of the spreading infection still kills and topples trees, but the center of the "wound" begins to heal. Eventually the edge of the infection will pass out of the forest.

In the past, the forest openings caused by *P. wierii* helped shrubs, hardwoods, and cedars to persist in forests where they otherwise might have disappeared. Today, openings provided by logging operations provide some shrubs with the same opportunity. In a sense, logging is an ecological substitute for laminated root rot; however, logging does not replace the disease; it enhances it. Logging increases the number of young, even-aged, low-diversity stands in which *P. wierii* thrives. Furthermore, when viewed as a disease, logging is vastly more destructive than laminated root rot.

TUBER GIBBOSUM, AN AMERICAN TRUFFLE

Our last fungus, the Oregon white truffle, like the chanterelle, is mycorrhizal. But it is *not* a mushroom: Its fruiting bodies are formed underground.

A truffle (the mushroom, not the chocolate kind) is the fruiting body of a fungus. The Oregon white truffle grows under the soil surface, usually within the humus layer, and just above the mineral soil. Its mycelium forms mycorrhizal rootlets with Douglas-fir, as does the chanterelle. Its renowned cousins, black and white truffles, are mycorrhizal associates of European oaks.

In a culinary sense, a truffle is a condiment or spice, not a vegetable. Used in minute quantities, like garlic, its strong, cheeselike odor enhances the aroma of pasta or other dishes. In the forest, its strong odor serves another purpose: attracting the attention of deer, squirrels, and mice. They dig up the truffle, eat it, and spread the spores around. For us a truffle is an expensive condiment; for many small mammals it is a wintertime staple. In Douglas-fir forests, truffles ripen during the winter when other food is scarce. It is hard to imagine that rodents such as flying squirrels would leave the relative safety of their treetops to dig for truffles, but the stomach content of trapped animals has been found to be nearly pure truffle in the winter. Truffles are highly nutritious compared with mushrooms.

When, following a gourmet dinner, flying squirrels or red-backed voles venture out into brushy clearings at the forest margin, they spread truffle spores in their feces and bury surplus truffles for future use. Since Douglas-fir requires a mycorrhizal partner for normal growth, these rodents' activity abet forest reestablishment by supplying fungus spores to tree seedlings.

In recent years, the Oregon white truffle has been discovered by local gourmets, and each year a few dozen pounds make their way to gourmet stores and restaurants in eastern cities. However, Oregon white truffles are not likely to compete with its European counterparts for some time, if ever.

We have considered six examples of fungi that live in ancient Douglas-fir forests. They are only representative of the thousands of kinds in these forests. Each of our examples can also be found in younger forests, but each is at home in the old-growth. What happens to the fungi when their home is destroyed; when the forest is cut, the valuable logs removed, and the site prepared for a new crop of trees?

AFTER THE FALL: FUNGI IN CLEARCUTS

When a stand of old-growth is logged, its fungal populations change profoundly. Most fungi disappear and do not return for many years, if ever. A few, especially those associated with brushy understory plants, may actually increase. Changes in the fungi reflect the changes in the site. The large trees are gone and with them all their associated fungi, including most of the lichens and fungi living in or on leaves and twigs.

Often the roots of large trees remain alive for a few years. The ecto-mycorrhizal mushrooms that depend on these living roots frequently produce a bumper crop of mushrooms in the year following a clearcut. Then they die.

Removing large trees opens the soil surface to the sky, resulting in warmer summer and colder winter temperatures. Bare soil is pounded by winter rain but dries out early in the summer. The site may be burned to get rid of unwanted branches. Many fungi that formerly decayed conifer needles and twigs disappear, in part because soil conditions have changed, in part because the supply of needles and twigs is reduced. A special group of weedy, opportunistic fungi appears briefly in new clear-cuts. Among these are the bright orange cup-fungus *Aleuria* and the soil lichen *Leptogium*.

Decay fungi in large logs or snags or in deep soil are partially pro-tected from the environmental extremes caused by logging. These fungi not only persist but probably increase in numbers. The supply of dead roots increases when the trees are cut, increasing the numbers of root rot fungi. If large cull logs and standing snags remain, those fungi that decay heartwood also persist. The interior of large logs is, like the deep soil, a relatively stable environment, less subject to the extremes of temperature and moisture than the adjacent soil.

Nobody knows exactly what percentage of fungi found in an ancient forest disappear when the forest is clearcut, but it probably exceeds 95 percent of all species. Some will reappear in the new forest that grows on the site, provided that a nearby old-growth forest supplies spores or other starters. Some will reappear at canopy closure in fifteen to twenty-five years. Others will return with the reestablishment of a mature forest soil with a deep litter layer. Still others, such as the lichen, *Lobaria*, are unlikely to be seen until the new forest is 100 years old.

In the past, natural openings in the forest caused by landslides, wind-thrown trees, or wildfire were surrounded by large areas of ancient forest, which supplied spores to "seed in" the fungi when a new forest was ready for them. Today, however, ancient forests are dwindling and young forests cover the hills. We hope that when these forests mature, the fungi will reappear. Undoubtedly many will, but others will have disappeared forever. □

The "Most Noble" Polypore Endangered

☐ James M. Trappe, Department of Forest Science, Oregon State University, Corvallis, Oreg., and USDA Forest Service, Retired.

Although many fungi are known only from one or a few collections, it is difficult to determine which are "rare and endangered." Knowledge of fungal distributions and abundance in North America is spotty. Some regions have been examined in detail over a long period; others have hardly been looked at. What is worse, many of the fungi are microscopic and hence not seen in the wild. The fleshy mushrooms are ephemeral; they appear, mature, and decompose in a few days or weeks. How does one know whether a species is rare or simply that its fruiting was missed by the observer?

A few fungi, especially the wood-decay species in the polypore family, produce perennial fruiting structures. Most of these are common and widespread. The largest and perhaps rarest of these is *Oxyporus nobilissimus*, literally the "most noble" *Oxyporus*.

According to the namer of this fungus, W. B. Cooke (1949), it was first discovered in the Cascade Mountains of Clackamas County, Oregon. Since then, only three or four additional finds have been reported: Lewis County and Mt. Rainier National Park in Washington and Linn County in Oregon. Despite the infrequency of these finds, *O. nobilissimus* is not inconspicuous: It is the largest of the North American fungi, perhaps the largest in the world. One specimen was 140 cm (4 ft. 7 in.) across and weighed about 136 kg (300 lb.) when fresh.

The upper surface of the fruiting body is covered with a thick, furlike growth of fibers, which are initially white but later become a deep, rich brown. The underside consists of tiny white tubes within which spores are produced. Each year *O. nobilissimus* produces a new layer of tubes; one can determine the age of the fruiting body by counting tube layers just as we count the annual rings of a tree. Specimens of more than twenty years are known, but no one knows the potential longevity of the species.

Oxyporus nobilissimus, an extremely rare tree fungus. *This modest-sized specimen was collected in 1983 by people who did not realize its rarity. It was the first to be found in some 40 years. A larger one, listed in the* Guinness Book of Records, *was 55 inches across and weighed 300 pounds. Only four or five have ever been found, all in ancient forests.*

O. nobilissimus seems to occur only with old-growth trees, primarily noble fir but occasionally with western hemlock. To have been found only four times, a fungus this conspicuous must indeed be rare. It is almost certainly more rare now than in the 1940s. The area of the first find was being clearcut at the time, so the *Oxyporus* almost surely was destroyed along with the trees. In fact, it is likely that only the Mt. Rainier collection site is still preserved.

Does a demonstrably rare and endangered fungus deserve protection as well as rare and endangered animals? It is equally part of the symphony of life found in the remnants of the vast old-growth forests that once covered the Pacific Northwest from the Cascades to the sea. Without the participation of fungi, those ecosystems could not have existed. Let us raise the clarion cry: "Save *Oxyporus nobilissimus*!" □

V

The Biological Values of Ancient Forests, Part 2

Genetic Diversity

Crop disease and pest species are generally more flexible genetically than are the higher organisms they attack. Monocultures—large acreages planted in a single crop variety—facilitate epidemics of diseases and pests, because the modern varieties we plant over extensive areas are strikingly genetically uniform.
(Margery L. Oldfield, *The Value of Conserving Genetic Resources* [1984])

The lowest level of biological diversity—genetic diversity—is a major reason for conserving ancient forests. People make little use of genetic differences among individuals within moss or songbird species. Genetic differences within tree species sought for timber are quite another matter.

In the following essay, Seri Rudolph, whose research on plant-animal interactions includes studies of insects in Douglas-firs, provides some insights on the importance of genetic diversity for Westside forestry.

128

Ancient Forests as Genetic Reserves for Forestry

□ Seri G. Rudolph, Department of Zoology, University of Washington, Seattle, Wash.

We hear much these days about loss of diversity in tropical rainforests. Current estimates are that the tropics are home to tens of millions of species, of which a large proportion will be lost before they are even described or named. On a single 100 m² plot of tropical rainforest in Costa Rica, researchers counted 236 plant species, many times more than a comparable plot of Westside forest. Is our land impoverished of biological diversity?

The answer is no; in fact, temperate forests are a wellspring of diversity *within* species. Studies in the last few decades have found temperate, wind-pollinated trees, especially conifers, to be among the most genetically variable organisms known. Current forestry practices are reducing this diversity both within and among populations. To better understand the options for preserving genetic diversity, we need to discuss how this diversity is organized and its value.

The resource value of the genetic variation within Douglas-fir (or any species) resides at three levels: (1) the presence of specific genes or combinations of genes which confer adaptation to local conditions, including microclimate, topography, and competitors; (2) the value of genetically based variation in and of itself as a strategic defense against pests and pathogens; and (3) the value of variation in allowing continued evolution in response to changing conditions.

Local adaptation has been demonstrated in Douglas-fir on a variety of scales. It should not be surprising that coastal and interior populations differ in cold hardiness, timing of bud burst, and response to moisture stress. But recent studies link genetic differences to different microclimates of north- and south-facing slopes at the same elevation, only two to six miles apart. Adaptive variation within and among populations is important to forestry because it forms the basis for predicting the performance of seedlings used artificially to regenerate a site after cutting. The Forest Service recognizes the importance of using locally adapted seed sources and has set up guidelines to ensure that both wild-collected seed and seed produced by selective breeding programs represent local stock.

This is a valued step but an insufficient one because it addresses only the first value of genetic diversity. Focusing exclusively on the adaptive value of specific genes is a little like choosing your furniture one piece at a time: You miss something by not considering the whole picture.

The possibility that genetic variability of forest trees might itself be an adaptive feature has only recently received serious attention. Trees in genetically uniform stands would be long-lived "sitting ducks" for insect pests and pathogens, which have much shorter generation times and can therefore evolve much faster. But insects and pathogens faced with an array of trees, each differing unpredictably from its neighbors in defensive chemistry, might have difficulty evolving all-purpose ways to exploit such variable resources. Although a local pest population might evolve a way to circumvent one type of defense, neighboring trees will be protected by different compounds at different concentrations, thus diminishing the chances of a general pest outbreak. This has been suggested as a reason why gypsy moths cause more damage in sugar maple plantations than in wild sugar maple populations. Patterns of damage to Douglas-fir by western spruce budworms are consistent with this interpretation as well. Genetic diversity provides a natural hedge against insect damage.

The final value of genetic variability, as a resource for continued evolution, is particularly topical as the world enters a period of unprecedented climatic change. As the Westside's climate becomes more and more like that of the Sunbelt, trees carefully bred to give optimal yield under 1989 conditions will become obsolete, but naturally variable populations may have a few genetic wild cards up their sleeves that will allow them to adapt and survive.

What is the best way to conserve the valuable genetic variability of our forest trees? In an intensive study, the California Gene Resource Program evaluated options for maintaining genetic diversity of Douglas-firs. It considered both in situ preservation (reserves or seed production areas) and off-site "storage facilities" (seed collections, clonal orchards, etc.). Although all of these approaches maintain genetic *variation* to some degree, only the in situ options maintain the *structure* of that variation and therefore the fine-scale adaptation and potential for coherent genetic change.

Although clonal orchards and off-site plantations facilitate selective breeding and research and are useful for conserving specific genetic material, they sample only a limited range of genetic variation and contain primarily trees with traits considered economically desirable at the time they were set up. The California Gene Resource Program (1982) concluded that in situ methods "are a more reliable means of maintaining variability at all levels until adequate research has been carried out." "Old growth virgin stands not previously harvested" are its first priority for preservation because they contain the most useful genetic variation.

One thing is certain: Our options for genetic conservation are narrowing as more and more acreage is replanted with nursery-grown seedlings. Planting with genetically "improved" trees could permanently foreclose the option of restoring site-adapted, native stock to a managed area. Although breeding broadly adapted, fast-growing trees for intensively managed sites makes sense in the current market, market conditions may change in the future. If such intensive management becomes economically unfeasible—perhaps because of rising costs for labor, pesticides, or fertilizer—the original "low-maintenance" site-adapted vegetation can be restored only if a genetic source has been maintained. As demand for different wood products shifts, different genetic traits or tree species could become preferable. This suggests the need to maintain a viable source of genetic variation that has not become degraded by domestication or inbreeding.

Old-growth, mixed-age stands may have a further advantage over even-aged second-growth, as pointed out in David Mulcahy's intriguing study (1975) of sugar maple. He analyzed leaf proteins of trees of different ages; since protein structure is coded in the DNA, differences among individuals in a particular leaf protein reflect differences in the genes they carry that code for that protein. Sugar maples that germinated in the same year shared similar proteins, but other proteins predominated in seedlings from other year-classes. This suggests that conditions in the year of germination were important in determining which seedlings survived and that these conditions selected for different characteristics in different years. Yet when mature trees of many year-classes coexisted and interbred at a site, the genetic variability of the population as a whole could be reconstituted. If this situation is common in forest trees, naturally reseeded even-aged stands (as well as planted stands) that are replacing older mixed-age forests could be narrowing genetic variation significantly each time a stand is logged.

Ancient forests, and the genetic resources they represent, are irreplaceable. We have only begun to discover the richness of local adaptation and the evolutionary potential they possess. Preserving these resources intact may well be the only way to preserve our forests and timber industries in an economically and ecologically uncertain future. □

Ecosystem Diversity

THE CONCEPT OF ECOSYSTEM SERVICES

Viewed from the distance of the moon, the astonishing thing about the earth . . . is that it is alive.
(Lewis Thomas, *The Lives of a Cell* [1974])

When astronauts journey into space, their lives are precarious. Their survival utterly depends on their propulsion systems, guidance systems, communications systems, and life-support systems. Leaving the Earth requires complex systems to provide the right amounts of water and oxygen, to eliminate carbon dioxide, and to recycle other wastes. But, fortunately, on Earth, if we do not overwhelm them, nature's life-support systems make engineered systems unnecessary.

Forests, as a major component of these life-support systems, provide free services essential to the Earth's habitability. They create soils on which civilizations depend and prevent their erosion. They regulate atmospheric composition by providing oxygen and removing carbon dioxide and various pollutants, cleaning the air and moderating the climate. They provide us with clean water and store and release it slowly, thereby minimizing both flooding and drought.

Forests are linked with other ecosystems in ways that are becoming clear to ecologists, geochemists, and climatologists. As yet, we do not know how much elimination of Brazilian rainforests will alter climate in Kansas and California. We do know that a growing fraction of Africa has been deforested, that Ethiopia has lost 90 percent of the plant cover it had in 1900, and that droughts and famines now recur there with agonizing regularity. We know that damage from the 1988 Bangladesh flood, the worst in its history, was greatly worsened by deforestation in northern India and Nepal. And we know that the consequences of eliminating the natural vegetation are not confined to other countries.

As human population growth increases demand for ecosystem services, we are destroying the ancient forests that provide them. What replaces ancient forests—cities, roads, farms, clearcuts, and tree plantations—are, at best, poor substitutes for what is lost. Amazingly enough, tree plantations do not even provide the knot-free, strong, rot-resistant heartwood that comes from ancient forests. Trees are renewable, as foresters often remind us, but under most current management systems, many of their products—rare species, high-quality framing lumber, beautiful scenery, and ecosystem services—are not.

The following sections provide examples of ecosystem services from ancient forests.

CREATING SOILS

. . . long before [humans] existed, the land was in fact regularly plowed, and still continues to be thus plowed by earthworms. It may be doubted whether there are many other animals which have played so important a part in the history of the world.
(Charles Darwin, *The Formation of Vegetable Mold Through the Action of Worms* [1881])

After climate, soils are the most important factor determining patterns of life in terrestrial ecosystems. Where soils are too thin, as on steep, cold mountain slopes (where they form slowly but are carried away quickly), few plants can withstand the violent buffeting of wind and rain. But soils are not just an anchoring medium. Their chemical composition and texture determine the availability of nutrients, water, and oxygen for plants.

Soils can have obvious effects on biological diversity. In Klamath and Siskiyou national forests, some slopes are cloaked with dense communities dominated by Douglas-firs and sugar pines, while nearby slopes of the same altitude and aspect support communities that have sparse vegetation dominated by widely scattered Jeffrey pines. The difference is that the latter soils are derived from serpentine, a rock with a chemical composition that inhibits growth of many plants.

Variation in soils also has less obvious effects. Two communities might have the same species, but the vegetation of the more fertile soil can recover faster after a disturbance. These less visible differences can have important ecological and economic consequences.

Living things do not only reflect soils; they create them. Soil is a mélange of the nonliving, the living, and the once-living. Along with the parent rock from which it is derived and the weather to which it is exposed, the life in and above the soil has major influence on soil properties.

Below the surface, out of sight, soil can teem with life. A square meter can have 2,000 earthworms, 40,000 insects, 120,000 mites, 120,000,000 nematodes, and extraordinarily large numbers of protozoa and bacteria, all of them moving, taking in food, releasing wastes, and reproducing, thereby affecting the soil's chemical composition and texture. Most soil organisms occur in duff (fine decaying organic matter) and organic-rich mineral soil just below it. And many of what we think of as epigeous (aboveground) species actually carry on most of their lives underground, such as fungi that send their fruiting bodies (mushrooms) above the surface only to disseminate their spores.

Plants penetrate upper soil layers in forest ecosystems with many square meters of root surface per m² of soil. Roots and their associated organisms break down parent rock by insinuating their way into cracks and surrounding rock fragments, then exuding chemicals that aid in extraction of nutrients. But roots of many plants, especially conifers, are not very effective at absorbing soil nutrients. Still, they get by with a little help from their friends.

The roots of most forest plants (including all Northwestern trees) associate with fungi to form mycorrhizae. Some mycorrhizal fungi have only one kind of host; others associate with many. Some plants have only one kind of fungal symbiont; others (such as Douglas-firs) can harbor many.

As true of all mutually beneficial relationships, the mutualism between plants and mycorrhizal fungi depends on their different abilities and needs. Plants are less capable than their fungal symbionts at absorbing phosphorus from soil. In return for receiving it, plant roots exude photosynthate—sugars and amino acids—that fungi cannot manufacture. Because mycorrhizal fungi are microscopic, difficult to culture in isolation, and live in an opaque medium, scientists have had difficulty unraveling their role in forest ecosystems. But it is clear that without them, trees would grow more slowly and many would not survive. As in many aspects of forests, the most important processes are not necessarily the most obvious.

After carbon, oxygen, and hydrogen, nitrogen is the element that plants need in the greatest quantity, and its scarcity often limits plant growth. As explained in Chapter 4, nitrogen gas is useless to plants unless it is incorporated into nitrogen-containing compounds, so some plants form underground mutualisms with bacteria to obtain nitrogen.

Some nitrogen-fixing bacteria live in nodules on the roots of a few early successional Westside species, particularly alders and *Ceanothus*. Others are scattered through the soil or inhabit large, decaying logs in ancient forests and postdisturbance stands where logs have not been removed or burned. Additional soil nitrogen comes from decay or leakage from nitrogen-fixing lichens in the canopy of old-growth stands.

Although nitrogen is crucial, a far larger biological contribution to forest soils is through photosynthetic fixation of carbon and its subsequent addition to soil through decomposition. Some plant biomass is eaten by animals and deposited on the forest floor, but most is never eaten. It does not, however, go to waste.

Here decomposers come to the fore. By breaking down animal wastes and the bodies of dead plants and animals, they convert organic compounds into CO_2 and H_2O. In the process, they release nutrient elements that the plants took up days, years, or even centuries before, making them available for uptake by plants.

Some decomposers are animals that eat wood. Although much of wood is cellulose and hemicellulose, complex carbohydrates that are difficult to digest, a few animals, such as termites, can break them down. Here, too, it is the termites' mutualistic association with other organisms that is central to their role in forest nutrient cycling. Termites eat wood particles and provide them to protozoans and bacteria in their guts. These microorganisms find termite guts congenial places to live and repay their insect hosts by sharing some of the sugars and nitrogen compounds from the wood they digest.

Other insects, including beetles and carpenter ants, bore holes in logs. Some beetle larvae feed on the part of the log richest in nitrogen compounds, the cambium layer between the bark and wood. Carpenter ants do not actually eat wood but excavate tunnels to house their colonies. In the process, they reduce the wood to tiny bits that are readily colonized by other decomposers. More important, tunnelers provide wood-decomposing fungi a means of entering the log and actively introduce many fungi.

Fungi are master decomposers. Lacking claws or jaws to rend wood fibers apart, they use a sophisticated arsenal of digestive enzymes. Some extract their nutrients from sound wood; succeeding fungi use progressively more decayed wood until only the least digestible compounds, the lignins, remain. These complex organic compounds last in the soil for many years, even centuries, before breaking down into CO_2. Lignins are important to soils because they glue soil particles together, retain just the right amount of water to support plant growth, and bind and release nutrients.

Thus, decomposers create the soil chemistry and texture needed for the growth of the next generation of plants.

MINIMIZING SOIL EROSION

What is left now [of the soils of Greece, is like] . . . the skeleton of a body wasted by disease; the rich, soft soil has been carried off.
(Plato, *Critias* [428–347 B.C.E.])

The force of a raindrop might not seem a threat to civilization, but legions of raindrops can dislodge legions of soil grains and send them roiling downhill. Ice, wind, and rain can undo the work of living systems by stripping soils from their birthplaces. The rainy and mountainous Westside would be especially vulnerable to soil erosion if living systems did not protect their investments. Fortunately, the mantle of vegetation provides a strategic defense against erosion.

As with any system of protection against aerial attack, this one has layered subsystems. The needles of the canopy intercept raindrops high above the ground. As drops amass and resume their earthward journeys, they are repeatedly slowed and stopped by foliage. Trees and shrubs rob much of their erosive force.

Drops falling from the canopy crash into mosses and decomposing plant material on the forest floor, expending their kinetic energy before reaching soil grains small enough to dislodge. The thick layers of moss and duff that accumulate in an ancient forest are the second layer of defense against erosion.

The third layer lies within the soil itself. Anastomosing plant roots and fungal threads bind the soil, just as steel reinforcing mesh strengthens concrete. On gently sloping lands, where sheet erosion and gully erosion are the major threats, this third layer is probably superfluous as long as the first two layers are intact.

But on steep slopes, where mass failure is the major source of soil erosion, roots play a crucial stabilizing role. Soils lose strength and become vulnerable to slumping or landslides when they become saturated with water. The network of roots often makes the difference between an intact slope and one that fails suddenly.

This is easiest to see where a steep site is logged and burned. At first, surface erosion increases because the protective canopy and duff layers are gone. Then, as tree roots decompose, the soil loses its cohesion and fails en masse. Mass erosion is common for many years after logging, after the original roots have rotted but before a new root network has replaced it. Until then, soil erosion can damage real estate values, hydroelectric dams, municipal water supplies, freshwater fisheries, and the biological diversity of stream ecosystems.

Storing Carbon

... it is obvious that climate exercises powerful constraints over the kinds and numbers of living things that can exist on earth. ... But what is probably much less well appreciated is the fact that life, as it multiplied and evolved over the aeons, altered the land, air and seas— enough to have changed markedly the very climatic conditions from which earlier life emerged. In a sense, climate and life grew up together, each exerting fundamental controlling influences on the other. (Stephen H. Schneider and Randi Londer, *The Coevolution of Climate and Life* [1984])

Whereas climate shapes the ecological processes on this planet, the connection goes both ways: Climatic patterns are affected substantially by living things. Plant communities affect the land's albedo (reflectivity), which determines how much solar energy heats the land and lower atmosphere instead of being reflected back into space. Transpiration by plants cools the air and increases humidity. But the most important way that ecosystems affect global climate depends on their rates of photosynthesis and respiration and, thus, the amount of carbon they store.

Plant photosynthesis removes carbon dioxide gas from the atmosphere and stores it as biomass in leaves, branches, tree trunks, snags, downed logs, duff, roots, and soil. Indeed, natural gas, oil, and coal are chemically altered biomass that has been stored in the Earth's crust for millions of years. CO_2 returns to the atmosphere when humans burn fossil fuel and trees and when living things respire. The balance between photosynthesis and respiration affects the amount of atmospheric CO_2, as explained below by R. A. Houghton, who has worked for many years on the global carbon cycle. CO_2 is the most important gas affecting the Earth's heat balance, the so-called "greenhouse effect."

Pacific Northwest Forests and the Global Carbon Cycle

□ R. A. Houghton, Woods Hole Research Center, Woods Hole, Mass.

The contribution of terrestrial ecosystems to the Earth's carbon cycle is substantial. The total amount of organic C in terrestrial vegetation and in the surface meter of soils is about 2,000 billion metric tons, almost three times the amount of C in the atmosphere (740 billion tons in 1988). The exchanges of C, as CO_2, between terrestrial ecosystems and the atmosphere are about 100 billion tons annually, potentially enough to double, or deplete, the atmospheric content of CO_2 in only seven to eight years. But because these photosynthetic and respiratory exchanges overlap in time, the actual seasonal exchange of CO_2 between terrestrial ecosystems and the atmosphere is on the order of only 10 billion tons of C for the entire Earth.

This seasonal exchange shows up in the regular oscillation of CO_2 concentrations observed at atmospheric monitoring stations throughout the world (Figure 5.1).

Forests are particularly important in the global carbon cycle because of the large amount of C they contain relative to other ecosystems. Forests and woodlands today cover approximately 38 percent of the world's land surface and contain in their vegetation and soils about 60 percent of the land's C. The Westside's ancient forests contain very large amounts of C per unit area relative to the world's other major forest types. The total amount of C (above- and belowground) in live vegetation of nonboreal coniferous forests averages about 168 tons/hectare. (Animals and microorganisms would add less than 1 percent.) In contrast, live vegetation in Westside old-growth forests averages 600–700

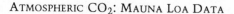

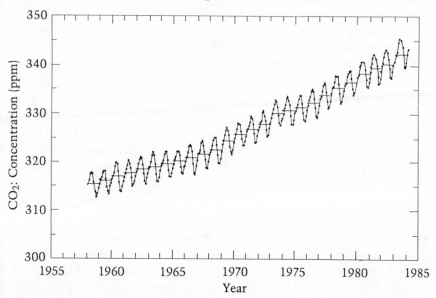

Figure 5–1. Concentration of atmospheric carbon dioxide on Mauna Loa, Hawaii. *This 30-year record shows two major patterns. First, the atmospheric concentration of carbon dioxide has increased by about 10 percent since 1958, when measurements began. This increase is caused by the burning of fossil fuels and by forest destruction. The second pattern is the regular, seasonal oscillation. Carbon dioxide concentrations increase each fall, when the respiration of living things exceeds plant photosynthesis, and they decrease in the spring, when photosynthesis takes up more carbon dioxide than respiration releases. This seasonal oscillation shows how living systems affect the global atmosphere. In a sense, it shows the Earth's breathing (from Gammon et al. 1985).*

tons of C/ha, with the highest measurements about 1,100 tons/ha. These averages are about three times larger than averages for tropical evergreen forests, although some tropical or temperate rainforests store much larger than average amounts.

In addition to their prodigious living biomass, Westside forests have the world's greatest reported amounts of downed logs and snags. In old-growth Douglas-fir/western hemlock forests, the amount of C in downed logs and snags averages 82 and 36 tons/ha, respectively, and can reach 245 and 53 tons/ha. In contrast, average values for logs and snags in old deciduous forests are only 15 percent and 7 percent as great, and coniferous forests in other regions are also much lower.

The organic carbon content of the soils in Westside forests is also large (averaging about 200 tons of C/ha, with some as high as 388 tons/ha), but not atypical of other carbon-rich soils such as in boreal coniferous forests. Taken together, the live vegetation, logs, snags, and soil carbon average some 900–1,000 tons/ha of C. No terrestrial ecosystems on Earth are known to have higher average values.

Before human activity, forests are estimated to have occupied 43 percent of the Earth's land surface and to have held about 27 percent more C than they now contain. Humans have decreased the forested area for millennia but probably never so rapidly as now. In 1980, the rate of deforestation in the tropics was about 11 million ha (about two-thirds of the area of Washington state) per year. Undoubtedly it has increased since then, although there has not been an updated comprehensive assessment. But it is clear that the rate in Brazil's Amazon rainforest is several times greater than in 1980. In 1987 the loss amounted to 6 to 8 million ha in this region alone. Outside the tropics, the rates of deforestation and reforestation are thought to be approximately in balance, with net deforestation in the United States and net reforestation in China and the USSR.

Deforestation has contributed to the increase in atmospheric CO_2. Carbon, as CO_2, is released to the atmosphere during deforestation when the organic C in trees and in soils is oxidized. When forests regrow, as, for example, following logging, the exchange is reversed: C is removed from the atmosphere and accumulated on land.

Despite the fact that regrowing forests accumulate more C than old-growth forests annually, the net effect of logging is very likely to be increased release of C to the atmosphere because C stocks in managed forests are much less than in virgin forest. One analysis suggested the average C stocks during a forest rotation to be one-third the stocks of the original forest. The fraction would probably be considerably less than one-third if the natural rotation length of 500–1,000 years were reduced to eighty or even fifty years, as is occurring increasingly in Westside forests. Further, managed forests have much smaller amounts of C stored in downed logs and snags. The fact that only a portion of the forest C ends up in long-lasting wooden structures, while the rest (roots, stumps, branches, leaves) is burned as slash, decays on the forest floor, or is oxidized in a few years (as paper, sawdust, other waste), also contributes a net release of C to the atmosphere.

Since the early nineteenth century, atmospheric CO_2 has increased about 25 percent. The major contributor in the last forty to fifty years is thought to have been the combustion of fossil fuels (coal, oil, and gas). About 20 percent of the current annual increase in the CO_2 concentration, however, and virtually all of the increase before 1850 were from deforestation. Current trends in fossil fuel combustion and deforestation are expected effectively to double the concentration of radiatively active gases (gases that contribute to the greenhouse effect) by the middle of the next century. This will cause global warming.

The rate of the warming is especially critical because rapid changes make appropriate societal responses more difficult. Although there are many scientific uncertainties about greenhouse warming, the potential hardships for society and the cumulative, open-ended nature of the warming suggest that steps should be taken now to stabilize concentrations of greenhouse gases. Increasing the efficiency of energy use could help; decreasing the rate of deforestation in the Pacific Northwest and tropical regions and increasing the area of forest through reforestation could also make significant contributions. □

Does Eliminating Ancient Forests Increase or Decrease Carbon Storage?

□ *Elliott A. Norse*

Since the searing summer of 1988, people have been speculating about the effects of Westside forestry on greenhouse warming. Is it better to preserve the ancient forests or to convert them to tree plantations? Some in the timber industry and the Forest Service have concluded that we should accelerate logging of ancient forests to help forestall greenhouse warming. Because Americans are finally seeing that global warming is a pressing issue, we need to examine this question carefully, lest we do the wrong thing.

Unfortunately, relevant data are scarce, but we can at least perform a mental experiment that suggests a rough answer and points to some gaps in our knowledge that must be filled before we can get a more rigorously quantitative answer.

It has been noted that young tree plantations accumulate carbon faster than ancient forests. This is probably true, but, oddly enough, it is not very relevant.

Obviously, a seedling carries on less photosynthesis than a huge tree, and an acre of nursery stock planted in a clearcut has far less photosynthetic tissue than an acre of ancient forest. But as saplings grow, their leaf tissue increases and they photosynthesize more each year. By the time a stand is three to five decades old, its amount of leaf tissue stabilizes and its rate of photosynthesis probably approximates that of an ancient forest. Presumably, then, accumulation of carbon is lower in a stand younger than about thirty years than it is in an old-growth stand.

However, as they age, trees acquire more nonphotosynthetic tissue that must be "fed" by photosynthesis. Their respiration increases while photosynthesis remains roughly constant. Hence, the amount of photosynthate they allocate to net growth gradually decreases. In the period from thirty years until rotation age (say, eighty years), a young stand undoubtedly accumulates more wood than an old one; Dean DeBell and Jerry Franklin (1987) suggest about twice as much. Ecosystem ecologists would say that their gross production is about the same but that net production in the young forest is higher.

But let's not start our chainsaws yet. Looking only at rates of wood growth is wrong for several reasons. One is that foresters measure only changes in aboveground biomass because they are most concerned with usable wood production. But as mentioned earlier, belowground production in a Douglas-fir stand equals aboveground production. Unfortunately, we know far less about belowground than aboveground patterns of production. Carbon accumulation may reach its maximum much faster or much slower. Because belowground production is so large and so dynamic, it should not be ignored in examining the ecosystem's C uptake.

In an ancient forest, trees also add large amounts of litter to the forest floor each year. Old trees add a lot of C to the ecosystem even if they do not make huge amounts of timber.

There is another, more fundamental problem with looking only at C fixation. The ecosystem's carbon balance is also determined by the various ways that fixed C is converted to atmospheric CO_2. And there is every reason to believe that old-growth and managed forests differ far more in the rates at which they lose C than in the rates at which they gain it.

When old trees die, they spend decades as snags or centuries as logs while their C oxidizes. Because decomposition is slow (recall that even a small twig takes perhaps a decade to disappear on the forest floor), huge amounts of C accumulate in vegetation, snags, logs, duff, and soil, more than in any terrestrial ecosystem on Earth.

When an ancient forest is logged, what happens to its carbon? In an old-growth Douglas-fir stand in the Oregon Cascades studied by Phil Sollins and coauthors (1980), 57 percent of the biomass (or C) in the ecosystem is aboveground vegetation. Subtracting foliage, twigs, branches, and stumps probably means that the boles that are removed account for no more than 50 percent of the C in the ecosystem. The rest is in foliage, twigs, branches, stumps, and unmerchantable cull logs (slash), snags, downed logs, litter (duff), roots, and soil. Slash burns oxidize them to varying degrees; pile-and-burns and hot broadcast burns can oxidize a very large fraction. I would also guess that more soil C (including roots, which constitute 20 percent of the live biomass in the system) is oxidized than added for at least a few decades after logging.

But what about the boles? Only about half of a bole (in other words, less than a quarter of the C in the ecosystem) becomes lumber or plywood. The rest—the bark and a lot of the wood—is burned for fuel or made into paper, which become CO_2 very quickly. The lumber survives as stored C for varying amounts of time. Perhaps old-growth lumber survives an average of half a century as door frames or beams before it is converted to CO_2. The weaker, less decay-resistant lumber from younger trees probably has a shorter average life span. The average life span of plywood is likely shorter still. So, the C in a bole oxidizes much faster than it would in an intact forest.

All in all, then, timber operations release a huge pulse of CO_2 in the few years after logging and result in storage of only a small fraction of C for a few decades. Together, the amount of C stored in young growing trees and in lumber is a small fraction of the amount stored in an ancient forest. And if climatic change increases tree mortality from pests and fires, the disparity in C storage between vulnerable young tree plantations and resistant ancient forests will only increase.

So, even ignoring any values of ancient forests but their role in increasing or decreasing atmospheric CO_2, eliminating them and replacing them with tree plantations are clearly mistakes. The only thing as bad would be preventing reforestation of cutover lands. Cutting ancient forest to increase carbon storage is almost like trying to increase the number of words in a library by emptying the shelves of books repeatedly and then refilling them.

Before we destroy the ecosystems that are the world's carbon storage champions, it seems wise to develop well-thought-out, rigorously quantitative carbon budgets both for them and for what would replace them. □

CLEANING THE AIR

... the phenomenon of chemical fallout [from the air] is one of the factors making possible the growth of productive old-growth forest on mineral soils with a very low nutrient status.
(J. P. Kimmons, *Forest Ecology* [1987])

The world's ecosystems have sustained themselves for hundreds of millions of years on finite resources. The reason they can do so is remarkably simple but has profound implications: The waste products of some species are resources for others.

Bear feces are feasts for dung beetles and coprophilous fungi. The fetid flesh of a decaying grouse is coq au vin for dermestid beetles and blowflies. Oxygen, the waste product of green plants, is the breath of life for a vast array of species, including humans.

Human economic activities emit airborne wastes that erode marble, acidify lakes, and corrode our lungs. But if we avoid overwhelming forest ecosystems, they can take many of our wastes and use them as fertilizer. Oxidized compounds of nitrogen and sulfur from fossil fuel burning, for example, are major atmospheric pollutants. But nitrogen and sulfur are also components of proteins and so are essential nutrients.

Each hectare of ancient forest can have more than a billion conifer needles and goodly numbers of angiosperm leaves. Although primarily adapted for photosynthesis, these surfaces are also filters for collecting the thin aerial soup of dust and droplets, natural and anthropogenic. Young forests have shorter trees and (below thirty to fifty years) lower leaf area. As Dick Waring and William Schlesinger note, "Processes of impaction and absorption [of nutrients in rain, fog, and dry particles] are likely to be dependent on canopy leaf area, and thus may increase in importance through succession" (1985, p. 128). The uneven canopies of ancient forests create turbulence that further aids precipitation on the needles. An engineer familiar with fluid dynamics would be hard-pressed to design a passive system more efficient at trapping a highly dilute suspension of nutritious particles.

Ancient forests are efficient pollutant traps because they are on the upwind edge of North America, where levels of airborne nutrients are naturally low. For eons they have snared nutrients that survived passage across the Pacific Ocean and incorporated them into the forest's biogeo-chemical cycles. Now, as forests downwind of industrialized areas are succumbing from excessive airborne nitrogen and sulfur compounds, Westside ancient forests have an unequaled chance of surviving, grow-ing, and, in the process, cleansing the air.

CAPTURING FOG

When the well's dry, we know the worth of water.
(Benjamin Franklin, *Poor Richard's Almanac* [1746])

Ironically, in the Westside, the lushest region of the contiguous forty-eight states, summer drought poses a real problem: Precipitation is scarcest just when organisms (including people) need water most. But ancient forests have a means of acquiring water even when there is no rain for snow. They capture fog.

This might seem odd. Even meteorologists who measure rainfall to the nearest hundredth of an inch often ignore fog. But fog can be crit-ically important to plants and animals. Some of the driest regions on Earth—the Atacama Desert of Chile, the Namib Desert of south-western Africa, and the Canary Islands—can have years with no rain at all. They would be nearly lifeless had their species not evolved wondrous adaptations for capturing moisture from fogs that roll in from the oceans to the west. Fogs from the Pacific also contribute signifi-cantly to Westside forest ecosystems, especially during the summer. As in other foggy areas, evolution has equipped Westside trees to capture this precious resource.

Old-growth trees reach into moisture-laden clouds that hug Westside slopes, trapping fog droplets as they do pollutant particles. Compared with the canopies of even-age tree plantations, the canopies of old-

Droplets on Douglas-fir foliage, and waterfall, Willamette National Forest (Oregon). *A cloud-piercing ancient tree, with perhaps 60 million needles, having a surface area of one acre, filters countless fog droplets from the air. These collect on the tips of needles and fall to the ground. Fog drip can add 8 inches of precipitation to an ancient forest during the dry months, increasing streamflow when it is most needed. Clearcutting diminishes this vital source of water for decades or longer.*

growth forests, with their gaps and trees of different species and ages, are more uneven. This surface roughness can only enhance their efficiency as fog traps by creating turbulence that increases collisions between swirling fog droplets and sprays of conifer needles.

The ability of ancient forests to intercept fog seems to contradict the view that timber cutting is good for water users. Hydrologists have observed that clearcutting increases streamflow from watersheds in many forested regions of the United States. This happens for two reasons. First, some precipitation intercepted by vegetation evaporates; remove the vegetation and more water reaches the ground and can run off or recharge water tables. Second, plants extract water from the soil and transpire it into the atmosphere. Elimination of vegetation therefore increases total streamflow by several percentage points. Some foresters have tried to increase streamflow to lowland agriculture and municipalities by deforesting nearby mountains.

However, logging does not increase streamflow in all forest ecosystems, as Forest Service hydrologist Dennis Harr (1980) has learned. In an old-growth Douglas-fir/western hemlock forest in the Bull Run Municipal Watershed, which provides Portland's water supply, streamflow decreased slightly rather than increasing as expected when 25 percent of two small drainages were cut. Harr hypothesized that loss of fog drip diminished streamflow. During a later forty-week experiment, Harr (1982) found that net precipitation in the ancient forest was 1,739 mm (68 in.), 29 percent more than in eleven-year-old clearcuts. On an annual basis, the net precipitation in old-growth was estimated at 2,494 mm, 25 percent higher than in the clearcut watersheds, with the greatest difference (202 mm, or 44 percent) occurring in late spring and summer when water demand is highest. Fog drip contributed an estimated 35 percent of annual precipitation under the old-growth canopy.

These remarkable results explain the unexpected earlier streamflow measurements. Interception of fog by the ancient forest exceeded any gains from clearcutting. And the harvest of water by ancient forests is longlasting, while any gains from clearcutting should disappear in about 30–50 years when leaf area reaches its maximum. Perhaps most important, the 8 inches of added precipitation during spring and summer come when trees and people most need water. In contrast, where clearcutting increases water yields, it does so mainly in the wettest times, when it is least needed.

The water that old-growth forests harvest from the sky during dry summers is a resource that the Northwest can ill afford to lose.

LESSENING FLOODING AND DROUGHT

Can we ever have too much of a good thing?
(Miguel de Cervantes, *Don Quixote de la Mancha* [1605])

The mild Westside temperatures and cloud-snaring mountains create ideal conditions for giant conifers. And when large portions of forest watersheds are clearcut, conditions are ideal for flooding caused by rain-on-snow events, as explained by Steven Berris and Dennis Harr (1987). The Westside's prevailing airflow from the Pacific makes winter temperatures exceptionally moist and mild. Most precipitation in the lowlands is rain or fog, and most at high elevations comes as snow. But at intermediate elevations (350–1,100 m in the western Oregon Cascades), most snow falls at air temperatures near 0° C (32° F). As a result, relatively little energy is needed to start melting it, and these elevations are called the "transient snow zone."

The canopy of an ancient forest intercepts a major fraction of falling snow, and the large, stiff branches of older trees can hold its accumulated weight. Because snow on the foliage exposes a large surface area to the air, it melts quickly when the air warms above freezing. Most of the snow reaches the ground as meltwater or as wet clumps, so the snowpack is usually thin or absent in ancient forests.

In clearcuts, the surface area of snow exposed to the air is much lower, so it melts more slowly in contact with warm air. As a result, snow accumulation is always deeper than in ancient forest, up to nine times as deep when Berris and Harr (1987) sampled a clearcut and an adjacent old-growth stand. On one day, the clearcut had 377 mm (15 in.) more snow, setting the stage for a rain-on-snow event.

Should weather patterns shift, bringing heavy, warm rains, the deep snowpack in clearcuts can melt rapidly. The addition of several inches of water in the snow to the amount in the rainfall makes rain-on-snow events the cause of nearly all the Westside's serious floods.

Rain-on-snow events can turn small streams into raging torrents that cut their beds and banks and swell progressively larger rivers. In upper reaches, the scouring currents undercut trees along the banks and carry invertebrates, fishes, and salmon eggs downstream to their deaths. In lower reaches, bridges, roads, buildings, and people can be swept away in lowland flooding. Further, when floods finally abate, the heavier suspended sediments settle. Salmon spawning beds can be smothered under large volumes of silt.

Clearly, high winter streamflows can be damaging. But low summer streamflows can also be damaging. Richard Myren and Robert Ellis (1984) of the National Marine Fisheries Service have presented an interesting hypothesis about old-growth forests and summer streamflow. They point out that for a few years after clearcutting old-growth, transpiration is decreased, so streamflow rates are higher (except where fog drip is significant, as discussed above). But then transpiration increases rapidly in the fast-growing, second-growth stand to a level well beyond that in old-growth. This means that eliminating old-growth has short-term net benefits but net costs for a much longer period, several centuries or more, a situation reminiscent of its effect on the amount of forage for deer.

For species such as trout and young salmon that need cool water with lots of oxygen, higher summer streamflow helps to keep temperatures and dissolved oxygen levels within tolerable limits. It also slows the spread of trout and salmon diseases during unusually hot, dry periods, such as in the summer of 1987.

Higher flow during the dry season also means that municipalities will have larger supplies when water use is highest, rather than having excessive amounts when they need it least. In 1987, many Northwestern cities were forced to restrict water use, and some, including Seattle, were so pressed that water rationing was considered.

Large portions of watersheds that supply Seattle, Portland, and other cities have already been logged, and there is unrelenting pressure from the timber industry to cut still more. Further cutting could either increase or decrease total annual streamflow, but by diminishing summer water supplies, it could constrain anticipated economic growth or necessitate construction of yet more dams.

Scientific Research

Human subtlety ... will never devise an invention more beautiful, more simple or more direct than does nature, because in her inventions nothing is lacking, and nothing is superfluous.
(Leonardo da Vinci, *The Notebooks, Vol. I* [1508–18])

Although botanists, zoologists, and ecologists are a diverse lot, if asked what imaginary research apparatus they would want most, I suspect that most would answer, a time machine. Why? Because today's biological patterns don't tell the whole story of life on Earth any more than any particular page tells the whole story in a book.

Even if we could only look to the past (an assumption that Dougal Dixon [1981] did not accept in his fascinating book on life 50 million years after the extinction of humankind), a time machine would allow us to understand the present and even the future much better than we can now. What biologist would not leap at the opportunity to study the social behavior of dinosaurs, how those small East African apelike creatures hunted, or what America's virgin forest ecosystems were like?

Unfortunately, we may never answer crucially important questions because so much information is irretrievably lost. But we can still study the world's outstanding virgin coniferous forest. The ancient forests of the Pacific Northwest are a scientific world treasure.

Why do scientists need them? First and foremost, ancient forests are a reference standard, the baseline against which we can measure other systems. Their species and the old-growth ecosystems themselves have developed over eons. To understand their behavior, we must look at intact ancient forests for their evolutionary context.

Why not study tree plantations or arboreta instead? Because they resemble a natural forest no more than a few orcas in an aquarium resemble a wild orca population. Wild orcas respond to storms and interact with the species of their realm. They are not confined in concrete pools, do not eat frozen, vitamin-laced fish, wear cowboy hats on their heads, or jump through hoops. In the same way, ancient forests are shaped by natural processes, which makes them different from artificial forests.

First, natural forests are messy. To maintain neat, artificial forests, managers scarify, burn, plant, spray, thin, and cut. In doing so, they eliminate most resemblance to real forests, so artificial forests are not very useful for scientific study of natural processes.

Second, the oldest Westside tree plantations or botanical gardens are less than a century old. The oldest natural stands are a millennium or more and have grown where others like them arose and died. Through millennia of storms and fires, insect and disease outbreaks, they have sustained their diversity and productivity. There is little evidence that humans can do as well. We may have heavy equipment, fertilizers, pesticides, chainsaws, computer programs, and troops of consultants, but nothing they produce has stood the test of time. Faced with the demands of a fiber-hungry world, foresters need ancient forests as models of sustainable productivity.

Finally, ancient forests have higher genetic, species, and ecosystem diversity than artificial forests. Scientists need this diversity to learn the lessons that can come only from comparison.

Fortunately, the Forest Service and the BLM have begun to create a network of Westside natural forest reserves for scientific research, including the H. J. Andrews Experimental Forest in Oregon. The Andrews—one of only fifteen National Science Foundation Long-Term Ecological Research sites—has hosted the world's most concerted, coordinated research on coniferous forest ecosystems. There are also smaller research natural areas, some of which have ancient forest. Together, they constitute only a minuscule fraction of the Westside. The Andrews, for example, is less than 1 percent of Willamette National Forest. Many research natural areas are too small for ecosystem processes within them to operate as they once did, and there is question about their viability. Nonetheless, they are indispensable sites for scientists seeking to understand the workings of forests and how they are changing.

The scientific value of Westside ancient forests is not just regional, however; it is international. They are unique in being both essentially pristine yet relatively well known. Large areas of the Amazon and Zaire basins are still pristine but are almost unexplored by scientists. The forests of Germany and New Hampshire are well studied but extensively altered. When scientists want to understand how forests worked before they were irrevocably changed, we can still do it in ancient forests.

Ancient forests also provide unique opportunities to explore the past. Trees record the passage of time: Older ones (naturally) record more. Every tree is a history book, and skillful interpreters can read them. Dendrochronologists can decipher past climates by examining correlations in the widths of their rings. Fire ecologists can reconstruct stand histories by looking at "catfaces" on resistant old trees. Understanding past climates and fire regimes is vitally important to prepare us for the changes expected in years to come.

Ancient forests are also valuable to biomedical science. Oncologists seeking treatments for cancer are currently doing advanced clinical studies on taxol from Westside Pacific yews.

Even the threats to ancient forests can stimulate science. The fear that spotted owls are being driven to extinction has stimulated conservation biologists to push the frontiers of understanding about demographic modeling and the theory of minimum viable populations.

Thirty years ago, hardly anyone realized the scientific values of ancient forests. The H. J. Andrews Experimental Forest was used mainly to study the best ways to cut and extract old-growth timber. Spotted owls were virtually unstudied. Ecosystem science had not yet been born. Cancer was a death sentence. Paleoecology was in short pants. "Impact" meant "collision." And we believed that people talked about the weather but could not actually affect it.

All that has changed. Scientists are discovering the complex linkages between our actions, the composition of the atmosphere, and the health of our forests, and we need ancient forests to monitor global changes. With each disappearing acre, the scientific value of remaining ancient forests will increase, at least until somebody invents a time machine.

Tree Plantations and Ancient Forests

TREE PLANTATIONS ARE DIFFERENT

An unlearned carpenter of my acquaintance once said in my hearing: "There is very little difference between one man and another; but what little there is, is very important." This distinction seems to me to go to the root of the matter.
(William James, "The Importance of Individuals," *The Will to Believe* [1897])

It is said that the language of Inuit ("Eskimo") people has seventy-nine or 200 or some other large number of words for snow. These are not mere synonyms but depict states as different as fresh, soft, dry powder and old, crusty, granular corn snow. This rich terminology indicates the great importance of snow in Inuit lives and their sensitivity to its variations. Unfortunately, we have only one word for forest.

Ancient forests and younger stands can differ dramatically, but they can also be quite similar. Because most stands begin with disturbances that kill many trees, but not all, younger stands can share key structural elements of ancient forests: large living trees, snags, and downed logs. That is why many species abundant in ancient forests can also occur in lesser numbers in younger natural forests.

In contrast, similarities between natural forests (of any age) and "managed forests" (tree plantations) are superficial. They can have some of the same animals and trees, yet they are as different as dry powder and

Elliott A. Norse, top and bottom

Structure of ancient forest and young tree plantation. *The ancient forest in the top photograph is complex, with a deep canopy formed by huge trees, smaller shade-tolerant trees beneath, scattered shrubs and herbs, large snags and downed logs, and a high diversity of species. In contrast, the 60-year-old tree plantation in the bottom photograph is structurally simple, with only one well-developed layer of foliage (the canopy), no large logs or snags except the legacies from the ancient forest, and low species diversity.*

Tim Crosby, top; Elliott A. Norse, bottom

Floor of ancient forest and young tree plantation. *The floor of the ancient forest in the top photograph supports a lush carpet of diverse plants. In contrast, the floor of the tree plantation in the bottom photograph is almost barren, with few plants concealing the litter fallen from the canopy. The abundance and diversity of animals differ in much the same way.*

Elliott A. Norse, top and bottom

Canopy of ancient forest and young tree plantation. *The canopy of the ancient forest in the top photograph has scattered gaps that were left when giant trees died. These gaps allow sunlight to reach down toward the forest floor, promoting the growth of diverse understory trees, shrubs, and herbs. In contrast, the canopy of the tree plantation in the bottom photograph is so dense that very few plants can grow in the darkness below.*

corn snow. Use of the same term to describe them reflects how much people do not know about forests.

Natural forests are miracles of complexity. The processes that generate them produce a mosaic landscape. Natural forests are complex because the patterns of disturbance that befall them are complex and because their species have evolved diverse responses to disturbance.

In contrast, the traditionally trained silviculturist aims to simplify the forest, to channel the maximum amount of nutrients, water, and solar energy into the next cut of timber. He cleans up the diversity of age and size classes that are less efficient to cut, skid, process, and sell. He eliminates slow-growing and unsalable trees, underbrush, and any animals that might harm his crop. He replaces natural disorder with neat rows of carefully spaced, genetically uniform plantings of fast-growing Douglas-firs. He thins and fertilizes to maximize growth. He applies herbicides and insecticides and suppresses fires to protect his crop against the ravages of nature that must be fought and defeated.

With such different forces shaping them, it is not surprising that natural and managed forests are different. That they bear the same name reflects an inability to see the forest for the trees.

Natural forests (especially ancient forests) and tree plantations differ markedly in species composition, structure, and ecosystem functioning (Table 5.1). Not having benefited from the hand of humankind, old-growth forests are messy, complex, and wild. In contrast, tree plantations are as neat, simple, and tame as cornfields.

Because so many natural processes disrupt the artificial order of tree plantations, plantation managers must invest large amounts of energy, material, and labor to protect them from undergoing succession. This is not only expensive but has large costs that are not measured in dollars. How to lessen these costs and how to maintain biological diversity and ecosystem services while maintaining a healthy timber flow are challenges worth the efforts of the best forest scientists and managers.

RE-CREATING ANCIENT FORESTS

The only solid piece of scientific truth about which I feel totally confident is that we are profoundly ignorant about nature.
(Lewis Thomas, *The Lives of a Cell* [1974])

I grew up when movies from the 1930s and 1940s could be seen ad libitum on television. If any movie from that era epitomizes humans' attempts at mastery over life processes, it is the original *Frankenstein.* Five decades old, it is as thought-provoking as ever.

In *Frankenstein,* a physician uses bizarre devices and lightning bolts to create life from the dead. Unfortunately, he does not get quite what he had hoped, and he and a village suffer for it.

The most ironic thing about the film is not the way that Dr. Frankenstein's best-laid schemes go awry; many have explored that theme. Rather, it is that he tried to create life two decades before the structure of DNA was deduced, before molecular geneticists began probing the secrets of cells. Dr. Frankenstein tried to write the *Encyclopaedia Britannica* before knowing what letters are. Not only could scientists not create life then; we cannot do it now. No one can make a "simple" bacterium from scratch, let alone a spotted owl or a human, and we should not expect to do so anytime soon.

Ecosystems are enormously more complex than individuals. They comprise many individuals of many species whose interactions further increase their complexity. No human fully understands even a carpenter ant colony, let alone the ecosystem in which it lives.

In our understanding of ancient forests, scientists are now in much the same place that we were with human anatomy and physiology when *Frankenstein* was filmed: We have identified the major parts and have measured the ecosystem's pulse, metabolism, and digestion. We know that a diversity of large living trees is essential, and (climate permitting) we could probably grow them in most places if we had the patience to wait for centuries. We know that large snags and downed logs are essential, and we can "create" them with dynamite and chainsaws. For a few sites, we have fairly complete lists of a few groups, such as birds and plants, and we could try to coax them into colonizing a re-created forest. Still, we are (at best) many decades from knowing all their major components, connections, and feedback mechanisms, although we can venture some informed guesses.

As long as we have ancient forests as models, we can learn how to sustain biological diversity and productivity by mimicking their features in managed forests. But can forest management re-create an ancient forest ecosystem, something that was only defined with any accuracy in 1986? The Society of American Foresters says no. The Wildlife Society says no. The scientists who pioneered ecosystem research in old-growth say no. The naked clearcuts that managers repeatedly fail to reforest say, mutely, elegantly, No!

People who maintain that they can do so evidence hubris: a dangerous

Table 5.1

ATTRIBUTES OF ANCIENT FORESTS AND INTENSIVELY MANAGED
TREE PLANTATIONS

Attribute	Ancient Forest	Tree Plantation
Structure (land)		
Canopy	Uneven; many gaps	Even; dense
Tops	Often broken	Most unbroken
Cavities in trunks	Many	Few or absent
Height of dominant trees	Uneven; often taller	Even; often shorter
Girth of dominant trees	Uneven; greater	Even; smaller
Subcanopy trees	Various heights	Absent or small
Shrub layer	Uneven; dense in clumps	Even; often sparse or absent
Herb layer	Uneven	Even; often sparse
Moss layer	Uneven	Even; sometimes sparse
Epiphytes	Abundant on trunks and large branches	Sparse or absent
Perched soils	On large branches	Absent
Snags	Uneven; small to large	Even; few or no large snags
Logs	Uneven sizes; many large, many decay classes	Even; few or no large logs except possible remnants of natural forest
Carbon storage	Higher	Lower
Overall structure	Complex; multiple, indistinct layers; heterogeneous; much coarse woody debris	Simple; fewer but more distinct layers; homogeneous; little coarse woody debris
Microclimates (land)		
Light level at forest floor	Uneven; sunny in light gaps	Even; low
Snow depths on forest floor	Uneven; shallower	Even; deeper
Temperature and moisture on forest floor	Uneven	Even
Species diversity (land)		
Trees	Higher	Lower; often one species

Table 5.1 (*continued*)

Attribute	Ancient Forest	Tree Plantation
Understory plants	Higher	Lower
Animals	Higher	Lower
Fungi	Higher	Lower
Overall diversity	Higher	Lower
Nutrient pathways		
Nitrogen fixation by epiphytes and in logs	Exceeds atmospheric inputs	Little or none
Nitrogen from early successional trees or shrubs	High	Absent due to vegetation management
Overall nutrient flow	Cyclical; many sites of nutrient capture and storage	Linear; in absence of capture and storage sites, nutrients leak from system
Disturbance regime		
Wind damage	Individual trees	Individual losses rarer; large-scale blowdowns more common
Damage from insects and pathogenic fungi	Usually individual trees or small groups	Epidemics can affect whole stands or larger areas
Overall pattern of disturbance	More frequent; scattered trees	Less frequent; whole stands
Structure (streams)		
Logs in streams	Many; some large	Few; none large
Gradient	Uneven; stair-stepped; pools common	Even; riffle and channel habitats predominate
Sediments	Diverse, from silts to cobbles	Uniform
Overall habitat diversity	Higher	Lower
Species diversity (streams)		
Invertebrates	Higher	Lower
Salmonid fishes	Higher	Lower
Amphibians	Higher	Lower
Overall species diversity	Higher	Lower

overestimation of their abilities. It seems they never learned the lessons of the price of overreaching, from Greek tragedies to the *Exxon Valdez* disaster.

Can we ever re-create ancient forests? With knowledge far beyond what Dr. Frankenstein had, the answer could be: Yes, conceivably, someday, but not for a long, long time.

VI
Effects of
Timber Operations

Destruction, Fragmentation, and Simplification

DESTRUCTION

The fact that [the Willapa Hills in southwest Washington] once supported one of the greatest forests on earth is beside the point since that forest isn't there any more—it's gone to sunken ships, secondhand furniture, derelict buildings, and yellowed newsprint. These are devastated hills, doing their best to recover, to grow green things in time for the next devastation. A ravaged land, awaiting the next ravages.
(Robert Pyle, *Wintergreen: Rambles in a Ravaged Land* [1987])

Chimpanzees, our closest relatives, are perhaps the next most intelligent animals. Much of their behavior closely resembles our own. Some years ago, a researcher trying a variant on an old experiment showed just how close chimpanzees and people are. He placed a young chimpanzee

161

in a room having a banana suspended from the ceiling. Also in the room were several boxes. As most chimpanzees would, this one tried to reach the banana to no avail, then stacked the boxes and climbed them. But even atop the stack, it could not quite reach the banana. Needing just one more box, it reached down to the one at the bottom and pulled it out, sending the whole stack crashing.

Humans, like the chimpanzee, often wind up undermining themselves. Few examples better illustrate this than the way we exploit forests.

Trees are renewable; they can replace themselves under the right conditions. So can forest ecosystems. In many (but far from all) cases, wherever natural forests now occur, those conditions are being met. When we meet those conditions, we should have infinitely renewable trees and forests.

As illustrated in Chapters 3, 4, and 5, the diversity in natural forests creates the conditions for continuing forest productivity. But the forestry being practiced on most private and even public lands aims to eliminate that diversity and concentrate resources (light, water, nutrients) into the one tree species that (presumably) will bring the highest profits when the stand is cut.

But no natural ecosystem has just one species, and natural processes tend to break up artificially simplified ecosystems, so they are inherently unstable unless we spend large amounts of energy, labor, and dollars in maintaining their simplicity. By managing against nature, we eliminate the processes that sustain forest productivity. In our haste to get the banana, we undermine ourselves.

Timber operations reduce biological diversity in three ways: by destroying, fragmenting, and simplifying ecosystems. Each phase—from road building to fire suppression—has one or more of these effects. And while some effects are obvious, others are subtle enough that scientists did not begin to understand them until the 1970s and do not fully understand them yet.

Timber operations have destroyed all but 13 percent of the ancient forest in western Washington and Oregon. The rest has been cut over and converted to housing, roads, croplands, clearcuts, and second-growth. Yet this underrepresents the actual loss because elimination of forest ecosystems has not been uniform. Some kinds of old-growth have been eliminated to a much greater degree than others.

Early loggers got logs to processing sites by floating them down large rivers. Understandably, they eliminated the nearby giant trees in riparian forests along rivers first. Then they plugged hundreds of smaller streams with splash dams, from which they released raging torrents to move logs. These activities radically altered the streams and riparian forests, erasing a rich component of biological diversity.

What were riparian forests like? In the deep, fertile, moist soils of floodplains and stream terraces, tree diversity was high, shrub and herb layers were very lush, and the risk of fire was much lower than in drier forests. Riparian forests were probably dominated by western redcedars larger than those currently existing, along with huge Sitka spruce (near the coast), Douglas-firs, western hemlocks, black cottonwoods, and (in some places) smaller broadleaf trees including red alders, bigleaf maples, and Oregon white ash.

Unfortunately, we probably will never again know anything like them. A few ancient floodplain rainforests remain in Olympic National Park, but more riparian forests were eliminated than any other type of Westside ancient forest.

After riparian forests were logged, the same economic and technological forces shifted timber operations to other kinds of lower elevation old-growth. Trees there were larger and easier to cut and transport than those in the mountains; loggers maximized profits by taking the best first. For example, Larry Harris (1984) of the University of Florida points out that Westside foresters earlier in this century distinguished "large old-growth" (most of it privately owned) from "small old-growth" (mostly in the national forests). Foresters have since dropped use of the symbols for the best-stocked categories of Douglas-fir stands with the biggest trees, apparently because they are no longer needed. Similarly, we are not likely to see 385-foot-tall or 16-foot-wide Douglas-firs again. As old records and photographs indicate, the ancient forests that remain—the small old-growth forests—must be a pale reminder of those logged earlier this century.

On public lands, the same economic forces have driven logging patterns. Destruction has been greatest at lower elevations; outside Olympic National Park little remains below 2,500 feet. The national forests are now eliminating old-growth in the highlands, where regeneration and regrowth are much slower. For example, as Harris notes, in Willamette National Forest only 10 percent of the cut in the first three decades of this century was above 4,000 feet, while 65 percent was by the 1970s.

Besides having ecosystems with the highest tree (especially broadleaf) diversity, biomass, and productivity, the low-elevation stands were also richest in vertebrates. In the Oregon Cascades, elevation is the most important variable determining diversity of amphibians, reptiles, and mammals (Figure 6.1). Birds are also more diverse in lower elevation forests, possibly in response to the greater diversity of broadleaf trees, which provide seeds and fruits uncommon in the pure conifer stands at higher elevations. Lower elevation forests also have larger populations.

Destruction of 87 percent of the old-growth and an even greater portion of riparian and low elevation old-growth has diminished ecosystem and species diversity. Timber operations also have more subtle effects: fragmenting and simplifying those ecosystems that are left uncut. These do not benefit biological diversity either.

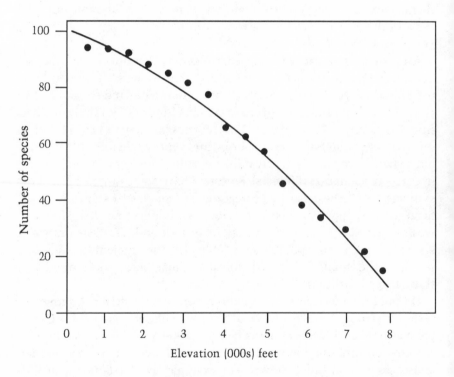

Figure 6–1.   Relationship between elevation and numbers of western Oregon *amphibian, reptile, and mammal species. Lowland forests, which have been cut to a much greater extent than forests at higher elevations, can support far more species. Leaving forests uncut at higher elevations does not compensate for the loss of lowland forests (from Harris 1984, p. 58; reprinted with permission, University of Chicago Press).*

FRAGMENTATION

The day is fast approaching when the remnants of the natural environ-
ment will be contained in a patchwork of parks and reserves. Much of
the world's biological endowment will then be locked into insular
refugia that are surrounded by an inhospitable landscape, through
which dispersal to the next refuge is slow or nonexistent.... we must
have large reserves in the size range of one to several thousand km² if we
are to prevent a veritable rush of extinctions.
(John Terborgh, "Preservation of Natural Diversity: The Problem of
Extinction Prone Species" [1974])

In the century and a half since Charles Darwin and Alfred Russel
Wallace left the British Isles to explore the biota of the Galapagos and
Indonesian archipelagoes, islands have had a special importance to the
science of ecology. Islands—large and small, near and remote, high and
low, old and new, inhabited by humans and uninhabited—provide comparisons of a kind that mainlands cannot.

In the 1960s, Robert MacArthur and Ed Wilson (1967) made a seminal
contribution to ecology and biogeography by analyzing species distributions on islands. They realized that the number of species represents a
balance between the number that reached and colonized the island and
the number on it that became extinct. They proposed a theory of island
biogeography that could be tested and used to predict things such as
extinction rates for species. Their work provoked and inspired many
other ecologists and had implications that go far beyond counts of
species on islands per se.

In the 1970s, John Terborgh (1974), Jared Diamond (1975), and other
ecologists examined similarities between islands surrounded by water
and islandlike fragments of once-continuous ecosystems and began to
look at ecosystem fragmentation and extinctions on mainlands.

In the 1980s, ecologists began to apply ideas from two rapidly developing disciplines, conservation biology and landscape ecology, to conserving species in fragmented landscapes. Larry Harris (1984) and Kenneth
Rosenberg and Marty Raphael (1986) began examining the effects of
fragmenting Westside old-growth forests.

Island biogeography has yielded insights crucial for understanding the consequences of fragmenting ancient forests. These include the following:

1. In many ways, ecosystem fragments *are* islands. Insularity means being surrounded by any barrier. To a species that does equally well in forests and grasslands, a forested patch in a sea of grass is not an island. But to one that cannot disperse across grassland, a patch of forest is as much an island as if it were surrounded by water.

2. Some species seem to colonize new fragments across unsuitable habitats poorly, *if at all*. Sometimes colonization is rare or nonexistent, so that more species occur in larger patches solely because they have lower extinction rates. This might seem surprising; it seems logical that organisms would have mechanisms allowing them to traverse unsuitable habitats. But as Mike Soule and coauthors (1988) found, even birds can have real problems dispersing among nearby habitat fragments.

3. Smaller islands retain fewer species than larger islands and for a shorter time. Species vary enormously in their need for resources such as food; a mouse's home range is thousands of times smaller than a bear's. Two groups—species that feed on widely dispersed or rare resources and species that need large amounts of space—cannot maintain breeding populations within small habitat fragments.

However, a fragment may not have a species even if it is large enough to support a small breeding population. Small populations are more vulnerable to extinction from random factors: adverse environmental conditions (e.g., droughts), population fluctuations, and fixation of deleterious genes and loss of beneficial ones. As a result, small fragments have higher extinction rates than large ones. But if anything, colonization rates are lower. If a species that vanishes from a small fragment cannot recolonize it, the number of species in that fragment declines.

Moreover, some species suffer when their ecosystems are fragmented because conditions change in ways that are anything but random. Compared with clearcuts, forests are cooler in hot weather, moister and less windy. Further, forest interiors are safe havens from species that live in edge habitats. There is evidence suggesting that in forest interiors, songbirds suffer less brood parasitism by brown-headed cowbirds and spotted owls suffer less predation by great horned owls than at forest edges. But the smaller the fragment, the larger the portion that functions as edge. The influence of external forces extends far enough that many fragments in today's landscapes are, in effect, all edge. For all these reasons, a fragment that now supports species from the original forest will not necessarily do so in the future.

4. Heterogeneous islands retain more species than homogeneous ones. The more different environmental features in a fragment, the more species it can retain. For example, a forest fragment having a permanent stream will retain more species than one without; a fragment having both moister (e.g., northeast) and drier (e.g., southwest) aspects will retain more than one with uniform exposure.

5. The effect of island size differs among organisms. Although all groups of organisms are more diverse on larger islands than on smaller ones, the effect of island size affects groups differently. For example (Figure 6.2), land birds in the Bismarck Archipelago (north of New Guinea) are more diverse on small islands than are reptiles and amphibians on small Caribbean islands, but bird species diversity increases more slowly with increasing island size. Fragmentation is likely to affect some taxonomic groups more than others.

All over the world, more and more species are in effect *becoming* island species as their habitats are being fragmented. Pacific Northwest rivers (many of which have been dammed) and ancient forests (which have been cut extensively) are no exceptions.

If you fly over the "checkerboard lands" of Mt. Baker-Snoqualmie National Forest, it is not difficult to tell which square miles are privately owned and which are managed by the Forest Service. Timber industry owners usually cut from one end to the other without missing a tree. The Forest Service requires loggers on national forest land to clearcut in a "staggered-setting" pattern of 10–20 hectare (25–50 a.) patches. Some implications of an idealized version of this pattern are explored in a groundbreaking paper by Jerry Franklin and Richard Forman (1987).

After enough holes have been cut from the forest, the landscape resembles an Emmenthaler Swiss cheese. But the cutting continues until the holes coalesce and the remaining natural forest becomes an archipelago. As islands are cut, the average distance between remaining ones increases. They become more isolated.

This process increases the amount of abrupt forest edge. As mentioned earlier, some foresters have seized upon the dogma that edges are "good" for "wildlife" and that staggered-setting clearcutting increases biological diversity, thereby justifying more logging. This ignores the fact that the increase results from colonization by opportunistic (weedy)

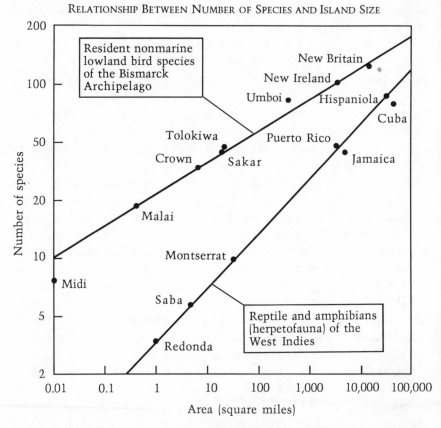

Figure 6–2. Relationship between number of species and island size. *Smaller islands have fewer bird and herptile species than larger ones, but herptiles increase more than birds as island size increases. Fragmenting ancient forests is likely to harm some groups more than others (from Norse and McManus 1980, p. 45).*

species. Indeed, the proliferation of edge has quite the opposite effect, making the remaining forests more vulnerable in several ways.

One is increased wind damage. The abrupt edges of clearcuts and roads increase stress on trees whose trunks and root systems developed in less wind-stressed environments. Trees at edges are more likely to be toppled by winds they could have withstood when they were protected by adjacent trees. Franklin and Forman (1987) point out that 48 percent and 81 percent of two major blowdown episodes in the Bull Run watershed of Mt. Hood National Forest were adjacent to clearcuts and roads.

Then there is climate. Increasing edge exposes forest interiors to harsher climates. There has been almost no study of microclimates in Westside ancient forests, but results from the southern Appalachians illustrate the point. Tim Seastedt and Dac Crossley (1981) of the University of Georgia found that summer temperatures at the soil surface averaged 26° C (79° F) within forests but 42° C (108° F) in adjacent clearcuts. If clearcutting warms the soil only half as much in the Westside, the hot, dry summer winds will still lick moisture from the lush forests.

Some ostensible benefits of edges have their dark sides. As ancient forests retreat, the clearcuts that replace them produce more forage for black-tailed deer and Roosevelt elk, spurring booming populations. But during snows, forage in clearcuts is covered over, and clearcuts do not offer hiding and thermal shelter as ancient forests do. Intensively managed tree plantations offer some thermal cover, but very little forage. Hence, where few ancient forest fragments are left, deer crowd into them and browsing pressure becomes severe. To the trained eye of a wildlife biologist, shrubs browsed to nubbins, tree boughs gnawed to tip-toe height, and saplings stripped of bark have an unmistakable meaning: poor management.

As scientists learn about ecosystem fragmentation, it becomes steadily clearer that the abrupt edges incising our ancient forests are no more beneficial than the hard edges of dams that fragment our rivers. To conquer nature, humans divide it. To live within and enjoy its benefits, we must protect and rebuild nature's connections.

SIMPLIFICATION

Chaos often breeds life, when order breeds habit. . . . Simplicity is the most deceitful mistress that ever betrayed man.
(Henry Adams, *The Education of Henry Adams* [1907])

Imagine an ecosystem that is totally flat: no hills, no valleys, no trees. The summer sun shines fiercely, the winter winds whip unobstructed, there's no place to hide. Imagine another, where the land is crossed by cliffs and streambanks, cloaked with trees of all sizes, standing and fallen, living and dead. Which would have more species?

The answer is obvious. The structural complexity of an ecosystem is a key factor determining its species diversity. As Robert MacArthur and John MacArthur (1961) pointed out, ecosystems with more three-

Peter H. Morrison, top; Elliott A. Norse, bottom

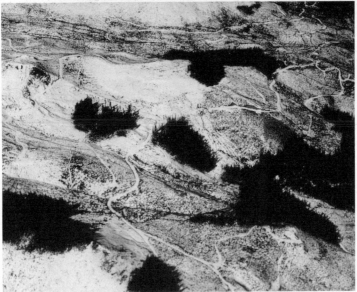

Forest fragmentation, Gifford Pinchot National Forest, with Mt. St. Helens in the background (top), and Mt. Baker-Snoqualmie National Forest (bottom) (Washington). *Dispersed clearcutting produces a landscape resembling a Swiss cheese. Further logging turns forests into islandlike fragments, which alters their microclimates and isolates populations, leading to extinctions. Because forest fragmentation is occurring even faster than forest destruction, it is the greatest short-term threat to ancient forests.*

dimensional structure have more species. A flat expanse of rock will support only species that can withstand exposure to heat, cold, rain, and wind and can deter their enemies without having to hide. But adding some boulders or trees creates diverse microclimates and refuges from stress and predators. Structural diversity provides opportunities for species that need vertical surfaces, horizontal surfaces, tangles, cavities, mating sites, and observation posts. Increasing structural diversity increases species diversity.

For example, a forest having a canopy but no shrub layer cannot support a hypothetical bird that forages in the canopy but nests in shrubs. A forest without streams cannot support aquatic and moisture-loving riparian species. A forest lacking trees with deeply furrowed bark will not provide as many hiding spaces for small creatures. A forest without snags and downed logs will not provide homes for many ancient forest plants, animals, and fungi. Complex habitats accommodate more species because they create more ways for species to survive.

With their deep crowns and diverse layers of understory trees; shrub-filled light gaps interspersed with densely shaded areas; furrowed bark and soil-covered branches; broken tops and epiphytes; and healthy trees, sick trees, and dead trees of different species—old-growth forests have exceptional structural complexity. That is why simplifying their structure by replacing them with clearcuts and second-growth reduces populations of species or eliminates so many species.

Much of the structural complexity in a new clearcut comes from any snags, stumps, and logs that were not removed or consumed by slash burning. They provide refuges from predators and harsh conditions. On warm, dry slopes, for example, seedlings are more likely to survive on the north side of a log, where they get some shade. Without the moderating effects of the canopy and coarse woody debris, clearcuts are harsh enough to exclude many species.

When the canopy closes, structural complexity falls sharply because most plants that cannot attain tree height get shaded and die. The young planted conifers have similar shapes, lack stiff branches that hold perched soils and epiphytes, and provide no large, furrowed trunks, no large snags or logs, no cavities. The diverse species of the ancient forest, replaced by the mostly different but diverse species in the clearcut, are then replaced by impoverished communities that tolerate low light and do not require the structures in ancient forests. Only after centuries, as structural complexity increases, would species diversity approach original levels.

Unfortunately, most forest planners are taught that multicentury rotations are uneconomical, so managed forests are cut long before they return to the diverse old-growth stage. Moreover, to maximize economic returns, the diversity of the early successional stage is also eliminated through vegetation management. The loss of structural complexity in these stages means fewer species and loss of functions that maintain forest productivity.

In addition to reducing structural diversity, timber operations simplify ecosystems in another way: by reducing the diversity of food resources. Replacing an ancient forest with a tree plantation monoculture eliminates many sources of food for animals. For example, the most commonly planted species, Douglas-fir, provides good cone crops only every five to seven years. The number of seeds available to birds (such as white crossbills) and mammals (such as Douglas' squirrels) in Douglas-fir plantations is low in most years. But mixed forests provide seed eaters with more constant food resources because other species (e.g., western hemlocks) produce seeds more reliably or in annual rhythms different from those of the Douglas-fir.

The higher diversity of food resources in ancient forests also includes fruits and insects. The fruits of Pacific yews, Oregon-grapes, huckleberries, and salmonberries, which are eaten by many Westside ancient forest animals, are rare or absent in tree plantations. In addition, the species-poor assemblage of insects in a tree plantation may have huge numbers at one time and very few the next, a boom-and-bust pattern providing an undependable food supply for insect-feeding birds, spiders, and insects. In contrast, the diverse food plants in ancient forests provide resources for many more insect species, whose population cycles are less likely to be synchronized. Availability of diverse seeds, fruits, and insects allows more kinds of animals to inhabit ancient forests. Diversity begets diversity.

For all these reasons, to the extent that timber operations destroy, fragment, and simplify ancient forest ecosystems, they reduce biological diversity.

Phases of Timber Operations

ROADING

Perhaps the most widespread pollutant of streams today is silt. Although silt is not toxic itself, it is still lethal. The adverse effects of silt

on aquatic systems have long been recognized by biologists, but little has been done to correct increasing siltation of streams.
(R. David Ono, James D. Williams, and Anne Wagner, *Vanishing Fishes of North America* [1983])

Timber operations occur in phases. Logging roads are the phase that makes all timber operations possible. Although they cost large sums to build and maintain, they admirably serve the timber industry's needs. Unfortunately, in the process, they damage streams and forests.

When Westside logging began, navigable rivers and streams provided the only access to timber. Mills and logging shows concentrated along their banks. After eliminating the riparian forests, loggers built skid roads for oxen (and later for horses) to haul out the huge logs. By the late 1800s, the greater muscle of steam locomotives allowed loggers to penetrate forests that were steeper and farther from mills. But no technology was so instrumental in eliminating ancient forests as the logging trucks developed since World War II.

Although later logging locomotives were amazingly surefooted, they could not make the grades that trucks can. Trucks and roads can conquer slopes once left to ancient trees and mountain goats. Hardly any Westside commercial forestland is too steep for roading.

Logging roads affect the land in many ways. First, they consume an amazing amount of space. A road punched through a wilderness does not take all that much land. But the anastomosing system of logging roads needed to access all the commercial forestland is another matter. Logging roads are the industry's circulatory system, and, like our own cells, every area to be cut must be near arteries, veins, or capillaries: the major and minor logging roads.

As Rollin Geppert and coauthors (1984) explained in a valuable report for the Washington Department of Natural Resources, each square mile of commercial forest requires about 5 miles of logging road. Although the road surface itself may be only 10–15 feet wide, the right-of-way is generally cleared to a width of 40–80 feet. As a result, about 10 acres are deforested for every mile of road (an average of about 50 a./mi.^{2}, or 8 percent). Thus, no matter what silviculture does to increase wood production on adjacent unroaded land, roads eliminate production on 8 percent of the land. In the nearly 30 million acres of commercial forestlands in northwest California, western Oregon, and western Washington, an area requiring 233,000 miles (more than nine times around the Earth) of logging roads, the loss of land beneath the completed road

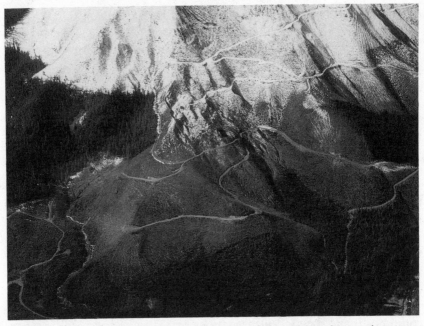

Elliott A. Norse

Logging roads, Mt. Baker-Snoqualmie National Forest (Washington). *Logging roads require the clearing of about 50 acres for each square mile of commercial forestland, eliminating about 8 percent of the forest. Logging roads increase surface erosion and landslides, increasing sedimentation in streams. They also fragment forests and provide entryways for fire, noxious weeds, insect pests, and diseases.*

system will amount to 2,400,000 acres, roughly twice the area of the state of Delaware.

But deforestation for roads and rights-of-way is only the beginning. Construction and maintenance of logging roads disturb large areas of soil. Most flat and gently sloping commercial forestlands (mainly privately owned) were roaded and cut long ago; nearly all those remaining (mainly in national forests) are steep, with slopes often exceeding 50 percent, even 100 percent (45°). Many cultures have learned the hard way that devegetating steep land strips the topsoil, sending it into streams and, ultimately, out to sea. Most of them are long gone.

Thus, logging roads accelerate the march of the mountains into the Pacific. On the way, the sediment smothers salmon spawning beds and clogs stream channels. Although sandbars in the Columbia River pre-date the timber industry, logging roads speed their formation. Fortunately, taxpayers from Bellingham to Miami are kind enough to subsidize the timber industry by funding the endless task of clearing sandbars from Westside rivers. But in smaller streams, sediment from logging roads just accumulates until storm discharges sweep them downstream.

Erosion does not end with road construction. Most logging road mileage is unpaved; unpaved roads produce hundreds of times more sediment than paved roads getting equal use. And logging roads carry heavy trucks, which damage roadbeds far more than light vehicles. Logging roads used by more than sixteen trucks per day produce 130 times as much sediment as roads restricted to passenger cars.

Surface erosion from roads under normal use leads to chronically elevated levels of sedimentation in streams during rainy periods, although, with time, sediment loss can diminish if the soils in roads become compacted. But when soils are saturated with water, a different kind of erosion, mass movement, or mass wasting (including slumping, debris avalanches, and debris torrents) can send huge pulses of sediment into streams. These "landslides" can cause far more erosion than the chronic surface erosion from roads, and their effects in streams are even greater than the quantity of sediment they produce.

Sediment from mass movements can scour the stream, eroding the bed and banks and eliminating stream biota. Downstream, they choke the feeding apparatus of animals such as the insect larvae that sift food particles from the water. And when the sediments settle, they bury living things, from caddis-fly larvae to salmon fry. The effects of scour and siltation can be felt for decades.

Mass movements can occur naturally in steep forested landscapes, but roading greatly increases their frequency. George Brown (1980) of Oregon State University noted that in one Willamette National Forest watershed, roads triggered forty-one times more debris torrents than intact forest; in another Oregon watershed, roads triggered 130 times more. Dick Fredriksen and Dennis Harr (1981) cite evidence that roads cause debris avalanches that increase erosion 25–340 times over that in intact forest. Many forest managers and scientists would agree with John Hewlett's conclusion: "Road and skid trail layout, construction, use and maintenance affect erosion from forest land more than all other forest activities combined" (1982, p. 149).

Logging roads have still other effects. During cold weather, traffic on logging roads disturbs wildlife such as deer, increasing their energy expenditures and hence risk of starvation. Roads increase the number of ignition sources, although they also provide access to fire-fighting crews. They spread noxious weeds such as tansy ragwort and spotted knapweed, insect pests such as gypsy moths, and diseases such as blackstain root disease. By increasing the abrupt forest edge, they increase the risk of blowdown. Snags are often eliminated near roads to protect road users, greatly reducing the number of wildlife nesting and roosting sites. And, as discussed earlier, logging roads fragment natural forests, isolating many forest interior species in islandlike enclaves.

CUTTING

I'd rather cut big trees, you know? Less cuts, more time on the saw. There aren't many old-growth left, though, to cut. But they're fun. . . . Good money in it, you know? And if you leave them standing they're just gonna go to waste, rot and fall down and they're worth nothing. (Ray Cokely, logger from Knappa, Oreg., *Oregon's Paradise: Lost?* KGW-TV, Portland, Oreg. [1988])

Of all phases of timber operations, cutting (or its consequences) is the most visible. Visual impacts of various cutting practices are a very real issue for a region that has offered some of the most spectacular scenery in the world. Ecological effects, however, are a different question. Clearly, forest destruction and fragmentation are diminishing populations of species that require the habitat elements most abundant in ancient forests. But what about cutting per se? Is it an unmitigated disaster as some environmentalists feel or a harmless practice as some timber managers assert?

The answer is not simple. Aside from the obvious destruction and fragmentation of natural forest ecosystems, the impact of cutting depends on where and how it is done and what else is done as part of the operation. Because Westside conifers do not resprout, cutting kills trees. In the process, it affects forest community structure, species composition, and hydrological and nutrient cycles. But the various cutting practices have different ecological effects.

Salvage cuts use various methods (usually clearcutting) to remove trees in areas that have been or are likely to be affected by fire, wind, volcanic eruption, or insects. It is often assumed that salvage cutting is harmless because these trees are dead and going to waste anyway. This neglects crucial facts: (1) that some trees taken in salvage cuts are alive and in no danger; (2) that snags and downed logs are crucial habitat elements for many species and important for maintaining forest productivity; (3) that punching in roads, cutting, and subsequent phases have important consequences for the recovery of the forest; and (4) that forests subjected to salvage logging are seldom allowed to return to something like their original condition but are turned into tree plantations instead.

Selection cuts remove single trees or small groups of trees, creating gaps not markedly larger than natural treefall gaps. Because they do not expose large patches of forest floor to direct sunlight and winds, they alter microclimate within the patches relatively little. This favors recolonization by plants that tolerate shaded, cooler, moister conditions, such as western hemlocks, but is not as favorable (at least in the less sunny parts of Washington and Oregon) for early seral plants that need higher light levels, such as Douglas-firs. If only the most desirable trees are cut, selection cutting changes the species and genetic composition of the forest. It also requires repeated entries, thereby requiring continued maintenance of roads. Selection cutting is little used in Westside forests.

The main objective of thinning cuts is to channel limiting resources—light, nutrients, or water—into the trees likely to bring the highest returns to loggers some years later. In most cases, temporary reductions in canopy cover are filled by the remaining trees. Thinning can be precommercial, in which thinned trees are not removed, or commercial, in which loggers market trees they cut. In single-species stands, thinning can resemble natural forest mortality, in that it usually removes weaker trees overtopped by faster-growing or older trees, but it often changes the species composition of the mixed forests typical of the Westside if managers eliminate "weed" and "defective" trees (ones whose lumber is less useful to timber companies). Unfortunately, these hardwoods and conifers can be the most useful trees to animals, plants, and fungi, help to maintain future site productivity, and slow the spread of insects and diseases. Multiple thinnings require maintenance of extensive road systems, with attendant adverse effects on forest ecosystems. Thinning is widespread in Westside forests.

Shelterwood cuts take thinning a step further. They remove most canopy trees but leave enough uncut to provide some shading to protect the soil from heat and drying, thereby encouraging reestablishment of seedlings. Some years later, when seedlings are well rooted and less vulnerable to soil drought, the remaining overstory trees are cut. On cool, moist sites, shelterwood cutting can favor western hemlocks or silver firs over the Douglas-firs whose timber brings higher prices. But by regulating the amount of shading, foresters can provide conditions

Elliott A. Norse

Recent shelterwood cut, Willamette National Forest (Oregon). *Shelterwood cuts provide some shade, thereby protecting regenerating seedlings or planted stock from drying. This is especially important on warm, dry sites, such as south-facing slopes. The sheltering trees are usually cut after a few years, but some forward-looking foresters are leaving them, as shown here. By allowing enough large, wind-firm trees to remain after logging, forest managers provide future snags for the many species of wildlife that require them.*

ideally suited for establishing desired species on sites that face south, on droughty soils, and in hotter, drier areas. There are other benefits: The remaining trees lessen erosion and heating of streams and are used by at least some of the original forest species. In the Blue River Ranger District of Willamette National Forest, managers are experimenting with leaving wind-firm living trees after the site is restocked. This uneven-aged system could be good for wildlife while still providing substantial amounts of timber. Shelterwood cutting is less common than clearcutting, especially in less droughty areas.

Clearcutting is the predominant silvicultural system throughout most of the Westside. In the short term, it minimizes problems for timber operators: They save the cost of selecting trees to be cut and don't have to exert caution to avoid damaging trees that will not be cut. And clearcutting per se requires only one entry, which minimizes the spread of some diseases and allows forest managers to do less road maintenance, or to close or even abandon roads, at least temporarily. (However, when clearcutting is a prelude to other intensive operations, the usual case, it requires many entries.)

But clearcutting radically changes the forest environment. It destroys the trees that forest species depend on. It decreases the interception of snow and increases the snowpack, increasing the risk of flooding during rain-on-snow events and covering the food sources of elk and deer. It increases wind fetch, the distance winds can sweep unimpeded before encountering an obstacle. The combination of long fetches and the abrupt edges of remaining patches in clearcut landscapes increases risk of blowdown.

These effects are compounded by another consequence of clearcutting: erosion. Loss of the canopy exposes naked soils to rain, increasing sheet and gully erosion. The roots of the dead trees decompose, decreasing their tensile strength within the soil. On steep sites, mass slumping increases. George Brown (1980) cites studies in two Oregon watersheds where clearcutting increased debris torrents by five to eleven times over their occurrence in intact forest. Mike Anderson and Craig Gehrke (1988) of The Wilderness Society point out that landslides from clearcuts were eight times as common as those from natural causes after a severe storm in the Mapleton Ranger District of Siuslaw National Forest; indeed, clearcuts produced more mass movements than roads.

Clearcutting changes the hydrological and sedimentary regimes of streams. While low gradient stretches can be choked in eroded sediments, other stretches can become sluices from which floods sweep cobbles, sediments, and logs. And while peak winter flows of headwater streams in clearcut watersheds can be higher than in intact forest, summer minimum flows can decrease. Many a logger can recall eating a sandwich by a seep or stream in a forest he had clearcut, only to find that the water had stopped flowing for years to follow.

Not surprisingly, species that live in trees or those that need shelter and moderate climate cannot survive the harsh conditions of clearcuts. Deprived of shade, streams warm during the summer, sometimes approaching or exceeding the temperature limits of cool-water species from insects to trout.

Clearcut and mass failure (slump), Olympic National Forest (Washington) (top), and sluiced-out streambed in a clearcut, Willamette National Forest (Oregon) (bottom). *The roots of trees bind the soil, increasing its strength. In the years after clearcutting, these roots decompose and lose their strength before those of regenerating trees can bind the soil again. On steep slopes, mass erosion, or landslides, including slumping and debris flows, is a common result. Intensive timber operations increase runoff during heavy rains and eliminate large logs in streams. Together, these can lead to unimpeded winter streamflow that can sweep gravel spawning beds, salmonid eggs, and fry downstream.*

Clearcutting can have serious, long-lasting effects on stream amphibians. Steve Corn and Bruce Bury (1989) of the U.S. Fish and Wildlife Service compared the presence of four species in Oregon Coast Range headwater streams in forests logged fourteen to forty years earlier with those in headwaters in unlogged forests. All showed clear effects of logging, as Table 6.1 demonstrates. Densities in logged streams ranged from one-half (tailed frogs) down to only one-seventh of those in unlogged streams (Olympic salamanders). Fine sediments were more common and had probably filled interstices between the rocks where these species take refuge.

Table 6.1

PROPORTION OF STREAMS IN WHICH COAST RANGE AMPHIBIANS OCCUR

Amphibian	Unlogged	Logged
Pacific giant salamander	100	70
Olympic salamander	61	15
Dunn's salamander	83	35
Tailed frog	96	35

SOURCE: Corn and Bury (1989).

Clearcutting is not disastrous for all species. Some fishes and amphibia can benefit from the higher light levels. Deer and elk find increased forage in regenerating clearcuts, and many widespread species of disturbed habitats, such as white-crowned sparrows and striped skunks, find clearcuts good habitats. These benefits last only a few years, however, ending when the canopy closes.

Clearcutting alters nutrient cycles. In the deciduous forests of Hubbard Brook, New Hampshire, Gene Likens, Herb Bormann, and colleagues (1977) found large nutrient losses occurring for years after clearcutting.

In an old-growth watershed in the Oregon Cascades that was clearcut and burned, Dick Fredriksen and Dennis Harr (1981) found that peak nitrate and ammonia losses in streamwater were about twenty-five times greater than before logging, and losses, although diminishing, persist fifteen years or more. However, Phil Sollins and F. M. McCorison (1981) found nitrogen loss to be very small compared with that in Hubbard Brook or the amount removed in the logs. Clearcutting increased nitrogen lost in streamwater from 0.8 to 2.0 kg/ha each year, less than 1 percent of the amounts lost from removing logs and burning slash. Similarly, L. K. Mann and coauthors (1988) recently found that nitrogen losses from leaching amount to only 1 to 31 percent of losses from biomass removal in two young Douglas-fir stands in western Washington.

Clearcutting is often justified as the silvicultural system most closely mimicking the stand replacement fires that are widely thought to be characteristic of the Pacific Northwest. A closer look suggests that the truth is not so simple. Clearcutting is always followed by removal of logs, which does not occur on naturally burned sites, where large amounts of snags and logs remain. It is often followed by site preparation, planting with selected stock and vegetation management. All of these differ from what occurs on natural burns, as discussed below. But the effects of clearcutting per se also differ from those of natural fires in many areas.

The common wisdom about the Westside is that fires are not common but are often holocausts when they occur, killing virtually all of the trees within the burn perimeter, much like a clearcut. For example, Richard Williamson and Asa Twombly said, "Historically, wildfires have burned large areas, giving rise to extensive even-aged stands" (1983, p. 9). For a decade or so, fire ecologists have recognized that the warmer, drier portions of the region have more frequent but (usually) less severe fires than the cooler, moister areas to the north.

But recent studies suggest that light burning is commoner than had been thought, which contradicts the idea that clearcutting mimics natural burns. For example, on dry sites in the Oregon Cascades, Joseph Means (1982) of the Forest Service found that fires return every century or so, perhaps more often. The regime of frequent fires of lower intensity is reflected in the age distribution of trees. Rather than even-age stands, these sites have trees of all ages, reflecting (1) the selective effects of less severe fires, which kill younger trees and species with thinner bark but leave many older ones with thicker bark; and (2) the capriciousness of fires, which can leave some patches severely burned and others nearby lightly scorched or untouched.

Even where there is ample evidence for large stand replacement fires, such as the ones that occurred some 500 years ago in central Oregon that were studied by Jerry Franklin and Miles Hemstrom (1981), the resulting stands can be anything but even-aged. Rather, trees colonize the burns over an interval of a century or more. The resulting stands are very different from the even-aged stands produced by clearcutting.

Studies by Peter Morrison (1984) in the Oregon Cascades suggest that smaller, less severe fires are actually common in some areas where stand replacement fires were considered the norm. Fires that eliminate forest canopy over large areas undoubtedly do occur at long intervals (centuries), but smaller fires appear to be more important in shaping the species and age composition of the forest than previously thought.

All these studies suggest that clearcutting does *not* closely mimic natural disturbances. As Means stated about dry sites in Willamette National Forest:

> Nature regenerates these stands in a manner resembling a partial cut and then establishes trees over the next 60–150 yr. This scenario has two main implications for management. First, shelterwood or selection silvicultural systems . . . will most closely mimic the natural system. . . . Continuous canopy cover will have beneficial effects on many dry sites. For example, the overstory will probably retard soil loss on steep sites with thin soils, and its shade may speed regeneration. (1982, p. 156)

Means added that clearcutting causes difficult regeneration problems in many hot, dry sites in Willamette National Forest and that some are still not successfully restocked after five or more plantings.

Despite its short-term economic advantages, clearcutting is far from environmentally benign. Its effects on forest-dependent wildlife and streams are severe and are not compensated by its benefits to widespread species that do well in open spaces. Nutrient loss following clearcutting per se seems less severe than in eastern deciduous forests, although losses from removal of logs, slash burning, and suppression of nitrogen fixers are anything but trivial. The difficulty in regenerating clearcuts on warm, dry sites suggests that clearcutting is not appropriate there. J. P. Kimmins (1987) of the University of British Columbia suggests that it is not appropriate for either dry or wet sites. To maintain the biological diversity and productivity of Westside forests, clearcutting should be used much more judiciously, if at all.

REMOVAL

Over the long haul of life on this planet, it is the ecologists, and not the bookkeepers of business, who are the ultimate accountants.
(Stewart L. Udall, speech to the Congress of Optimum Population and Environment [1970])

Picture an economic system that has producers and consumers, competitors and regulators, income, outflow, and capital, yet is not driven by flows and accumulations of paper currencies or rare elements (such as gold and silver). Rather, the important flows and accumulations are of common elements, such as carbon and nitrogen.

Such systems still exist. In fact, since long before Adam Smith and Karl Marx, through disaster and recovery, they have proven remarkably resilient. Neither capitalistic nor communistic, they are not run by theoreticians, bureaucrats, accountants, or lawyers. Although they form the basis of great wealth, their secrets are not revealed in *Forbes* or *Business Week*. Indeed, their basic economic principles are virtually unknown to most economists. They are natural ecosystems: coral reefs, marshes, tundras, prairies, and, yes, forests. Left to themselves, they operate according to what Charles Darwin called "the economy of nature."

Let us not be overwhelmed by the similarities. Ecological systems and economic systems have many, even to the common origin of "ecology" and "economics," both of which refer to the management of *oikos*, Greek for "house." But they have diverged considerably since their birth as organized fields of study. To an ecologist in a society dominated by economic precepts, their differences stand out sharply.

It is interesting to speculate on what would happen if it were not economists but ecologists who provided the central theorems in Washington, Moscow, Tokyo, and Lagos. Unfortunately, that requires more rumination than can fit in this book. But of one thing we can be certain: The flows of paper and precious metals would not be the only flows accounted for in decision making. This includes decisions about what happens after a forest is cut, steps that begin with removal.

Trees contain chemical elements essential to the growth of living things. Some are common; none is so rare as gold or silver. But it is astounding that the forestry being practiced on public lands pays so little attention to these elements in its economic calculations; it accounts for dollars far better than for nitrogen. One result is that in biogeochemical terms, silviculture usually operates at a deficit.

Capital is a pool of something valuable, such as dollars or kilograms of nitrogen, that can be enlarged by income and drawn down by outflow. In recent years, the U.S. government has spent a yearly average of $140 billion more than it received in revenues (about $600 per person per year). This has drawn down its capital and created pressures to "privatize" national assets by selling them to reduce the deficit. Thus, the federal government has sold the Conrail freight railroad system, gold coins, and record acreages of ancient forest.

Private interests will continue to operate many railroads they acquired, so society has not really lost them. Nor is the gold lost. Our government can always buy it back. But ancient forests are something else. When they are gone, they are gone. No amount of money can bring back their pristineness, their beauty, their species, or their nitrogen.

Because the limited supply of soil nitrogen often limits plant growth, anything that affects nitrogen flux—income to the soil or outflow from it—affects plant growth. Cut the income or raise the outflow and you diminish nitrogen capital. Removing biomass raises nitrogen outflow.

Plant parts have markedly different amounts of nitrogen. In an ancient Douglas-fir stand, the nitrogen concentration in needles is slightly more than 1 percent; boles, branches, and twigs (combined) average little more than 0.5 percent. So far, so good; the most widely used silvicultural practices in the Westside remove only boles (trunks and bark), which are low in nitrogen. But boles so outweigh the rest of the tree that their removal eliminates a large share of the nitrogen in forest biomass. Moreover, there is increasing pressure for "whole-tree harvesting," which removes branches, twigs, and foliage as well as boles.

Because these other parts have higher nutrient concentrations, whole-tree harvesting greatly increases nutrient loss. Mann and coauthors (1988) found that in fifty-five-year-old Douglas-fir plantations in western Washington on fertile and infertile soils, removing boles eliminated 478 kg/ha and 161 kg/ha of nitrogen, respectively. Whole-tree harvesting yielded 13 percent and 23 percent more biomass, but the additional biomass had nitrogen concentrations about four times as great, so it eliminated 52 percent and 102 percent more nitrogen than removing boles only. Whole-tree harvesting increased removal of phosphorous, potassium, and calcium in roughly similar ways.

Depletion of nutrient capital will not help future forest productivity. But removal is only one way that timber operations affect forest nutrient income and outflow. Later sections will examine some effects of other phases on the forest's nitrogen economy.

Removing logs—skidding and yarding—has other effects. Old-growth logs weighing many tons must be moved to landings where they can be loaded onto log trucks. Tracked vehicles are usually used on flat and gently sloping sites.

When logs are hauled across the land, the soil is scarified (denuded) and compressed. Denudation exposes soil to rain-splash erosion, a minor concern on flat sites. Indeed, scarification is considered beneficial where Douglas-firs will be seeded or allowed to regenerate naturally since the seeds grow best on bare mineral soil.

But soil compaction is another matter. Soil texture markedly affects productivity. Soil particles that form crumbly aggregates allow roots to grow and provide sufficient pore space for the oxygen that roots need. Soil compaction prevents roots from growing, diminishes pore space, and increases susceptibility to root pathogens. Soils can remain compacted for decades or more, substantially decreasing productivity. High lead, skyline, balloon, and helicopter removal compact soils less than tractors. The more that logs are lifted, the less that soil is compacted.

The effects of removal per se on biota does not seem to have been studied, except in terms of its effects on tree seedlings. Disturbing the duff layer undoubtedly harms some species but provides opportunities for others. Except for root pathogens, however, it is unlikely that soil compaction benefits any living things at all.

The first three phases of timber operations have major impact on the economy of the forest ecosystem. But there are more to come.

SITE PREPARATION

... we can destroy the labyrinthine structure of a forest soil in milliseconds with a bomb or in hours with a bulldozer, yet it will not be coaxed back again before decades of slow successional change have paved the way for its return.
(David Ehrenfeld, *The Arrogance of Humanism* [1978])

In the early days, logging ended with removal. "Cut-and-get-out" operations were unimpeded by notions of land stewardship. There is some consolation in knowing that the lands they cut—those with the biggest trees—were such good lands that nature usually provided a second-growth forest, sooner or later. But as timber companies acquired land, their ethic changed to a much more progressive one. Land-holding companies such as Weyerhaeuser knew that their future would depend on their effectiveness in growing more trees. They started developing intensive silvicultural systems to maximize profits in the next rotation. In a business where tree growth rates determine profit, they could not afford to leave regeneration to the vagaries of nature. Instead, they restocked their cutover lands.

On public lands—typically higher in elevation, colder, steeper, less productive, and less favorable for seedling establishment than private lands—regeneration often requires decades. Although it would be a wonderful world if all people felt responsible for restoring land after taking their profits, the timber companies did not do so. It is not difficult to understand why. As Garrett Hardin (1968) explained brilliantly in "The Tragedy of the Commons," private citizens who use resources held in common have little incentive to conserve them because investing in the future is likely to benefit someone else. Thus, while timber companies had a strong incentive to manage their own highly productive lands for maximum production in the next rotation, they had a strong *dis*incentive to care for the poorer lands that they did not own.

The incentive to restock public lands was weak for public officials too. When states and federal agencies seemed to have endless land and trees sold for pennies per thousand board feet, there was little pressure to ensure the continued productivity of public lands. But as the virgin forests disappeared, it became clear that natural succession did not maximize income to state or federal treasuries. Laws increasingly required reforestation. The National Forest Management Act now prohibits logging in national forests where trees cannot be reestablished within five years, although many sites that are logged are so difficult to restock that replantings fail repeatedly.

The first step in restocking is usually some sort of site preparation. Cutting and removal make the site more favorable to species (such as Douglas-firs) that germinate and grow best on exposed mineral soils at high light levels, but site preparation per se goes farther. It can involve deliberate mechanical soil scarification to remove duff and competing vegetation. It can involve pushing logging wastes into streams (a practice that governments now discourage) or pushing them into piles to decompose. But the most common form of site preparation west of the Cascades is slash burning.

Shrimpers use the term "trawl trash" for the 80–99 percent of animal biomass that they kill and discard. Miners use the term "overburden" to describe the soil and green stuff that lie atop an ore deposit. Such terms suggest that these are seen as problems to be disposed of, not valuable resources. Foresters use the term "slash" for stumps, unmerchantable logs, tops, branches, and foliage left after cutting and removal. They burn it for several reasons:

1. Slash and duff generally make poorer seedbeds for Douglas-firs than the mineral soil they cover.

2. Slash contains valuable nutrient elements. After burning, nutrients in the ash become available to plants faster than they would if slash were to undergo microbial decomposition.

3. Slash burning can be useful for controlling dwarf mistletoes, parasitic plants that might otherwise diminish future yields of western hemlocks or some other trees on the site.

4. Large accumulations of slash can be dangerous for people to walk on, especially on the steep sites so common in national forests.

5. Slash covering the soil can hamper restocking, so burning allows crews to replant the site.

6. Slash burning can prevent wildfires. Large accumulations of slash create high fuel loadings. Slash burning during cool, moist weather can remove enough fine fuels so that the site is much less flammable during hot, dry spells. Many devastating fires before the 1920s began in thick accumulations of slash on abandoned logging sites.

There are two kinds of slash burning: "pile-and-burn" (or "windrowing") and "broadcast burning." Piling usually employs tracked earth movers to concentrate slash, litter, duff, and topsoil in piles that cover 20 percent or less of the site before the material is set ablaze.

Piling is less expensive than broadcast burning because it does not require a contingent of people to ensure that fire doesn't escape into surrounding forest. Unfortunately, it is also considerably more damaging. As Fredriksen and Harr note, "Tractor piling increases soil disturbance and compaction often causing long term erosional impacts on the site and loss of soil productivity" (1981, p. 242). In the South Umpqua Experimental Forest, they report that tractor piling and burning exposes 89–95 percent of the soil and compacts 26–27 percent. Piling concentrates the ash, depriving the rest of the site of nutrients. And windrows burn at higher temperatures than broadcast burns, volatilizing more nitrogen, sulfur, and other nutrients. Aside from being less expensive in the short term, windrowing has nothing to recommend it.

Broadcast burning eliminates the slash, litter, and duff in situ. When properly done, it is far less damaging than windrowing, but it has adverse consequences as well. Nitrogen losses in the duff alone (not including a comparable amount in the slash) can be several hundred kg/ha, more than intensive management replaces during a 80–110-year rotation unless the site is fertilized.

Broadcast burning has other effects. Some soils become less permeable to water when subjected to hot fires, which increases runoff.

Soil scarification and elimination of duff and slash also expose soils to the direct effects of rain and wind. By denuding soils, site preparation increases rain splash and gully erosion on steep sites. Broadcast burning can sterilize the uppermost layers of the soil, although its effects on underground components of cutover areas are unclear. Soil is a good thermal insulator, and many roots, seeds, spores, and animals can survive cool burns if protected by a few inches of soil. Of course, aboveground parts of plants are killed, so broadcast burning precludes advance regeneration. Where fire is virtually absent (such as Sitka spruce/western hemlock forests) and organisms are not adapted to fire, broadcast burning can prevent the return of anything resembling a natural plant community. And it poses still another danger: Escaped broadcast burns are one of the most common causes of forest fire in

Elliott A. Norse

Edge of recent broadcast burn and dead trees, Olympic National Forest (Washington). *This clearcut was burned a few days before the photograph was taken. Trees at the edge of the burn (light color; healthy trees are darker) were killed by ground fire that baked their roots or by hot air from the burn that cooked their aboveground tissues. Escaped slash burns are one of the most common sources of forest fires in the Pacific Northwest.*

Washington and Oregon. Confining a burn during dry weather is tricky, and failure causes localized or even major losses of adjacent forest.

Like natural fires, broadcast burns stimulate sprouting and germination of fire-dependent plants (such as *Ceanothus* spp.) that some foresters consider competitors of conifers. As a result, burning often leads managers to still further manipulation: vegetation management.

Finally, burning hastens the conversion of fixed carbon in slash to atmospheric CO_2, an unwelcome addition to greenhouse warming.

And there is a cloud on the horizon. Broadcast burning generates smoke. As population grows, there is growing opposition to burning because it degrades air quality and aesthetics. As a result, many foresters are resigned to its eventual elimination as a means of site preparation.

REGENERATION WITH SELECTED TREES

Schlimmbesserung *(German): a so-called improvement that makes things worse (noun).*
(Howard Rheingold, *They Have a Word for It* [1988])

In Third World nations, logging often means permanent loss of forests because hungry people move onto cutover lands to grow crops until the soil gives out. In the Westside, most land that could be farmed was deforested long ago. But deforestation still occurs. Conversion of forestland to other uses—roads, houses, and commercial development—although not nearly as extensive as in the Third World, is proceeding quickly as the population grows. Even on lands that are not converted, foresters often have difficulty reforesting cutover lands.

Given that no one—from conservationists to the timber industry—wants to see the Westside deforested, it might seem that reforestation is, therefore, good. But most humans who attempt to re-create nature cannot resist the urge to try some improvements. Although Westside ancient forests produce some of the world's most valuable timber, silviculturists and economists want more.

And so we regenerate cutover lands our way. In the Westside, that usually means planting with selected Douglas-firs, occasionally noble firs, or other species. Seri Rudolph examines below some implications of artificial regeneration of Westside lands.

Regeneration: Genetic Improvement or Impoverishment?

☐ *Seri G. Rudolph, Department of Zoology, University of Washington, Seattle, Wash.*

Early efforts to hasten and direct the regrowth of timber on cutover lands took the form of reseeding with selected tree species. Little attention was paid to the geographic origin of these seeds: One Douglas-fir was like any other, wasn't it?

The answer to this question became obvious as worried foresters reaped a harvest of slow-growing, unproductive, often disease-ridden plantations from seeds and seedlings planted off their native sites. These experiences pointed out the enormous practical importance of understanding the patterns of geographic and genetic variation within forest tree species. They led to the passage in 1939 of the USDA Forest Seed Act, a set of guidelines for seed collection and transfer to ensure that logged sites are replanted with locally adapted trees. These guidelines still form the basis of regeneration practices.

What, exactly, is this local adaptation that forest geneticists strive to conserve? The variation we observe among living things has two primary sources: genes and the environment.

Environmentally controlled variation is not inherited; it reflects the unique history and experiences of each individual. For plants, these influences include the soil type in which the seed happens to germinate, sun exposure at the site, and events such as fire and storm damage. Variation due to genetic differences, however, resides in an organism's DNA and can be passed on from generation to generation.

If certain genes are associated with better survival and reproduction, trees carrying them will leave more offspring in the next generation. This increase of certain genes in the local population—the process of natural selection—produces adaptation to local conditions. The term "local" is a relative one. When traveling in Nepal, anyone you met from your home state might be welcomed as a "local"; at a town meeting, you would accord this status only to residents of your immediate neighborhood who shared your concerns over the same schools, streets, and parks. Likewise, plants can become locally adapted to conditions ranging from broad regional climatic types to microsite variations in soil type.

Later I will return to the implications of local adaptation on finer scales. Here, I focus on the intermediate-scale local adaptation recognized by the Seed Policy Act and associated regulations. These regulations are designed to prevent seed transfer between areas of dissimilar climate and elevation, mapped out in a system of tree seed zones. In California, seed zones range from 400 to 5,000 mi.^{2}.

These guidelines also address a major concern of conservation biologists: the loss of genetic diversity within populations. Genetic diversity (see Chapter 5) is critical to a species' ability to adapt to environmental change, and Douglas-fir has one of the highest levels of within-population genetic diversity of any species! The Forest Service cone collection guidelines specify minimum numbers of stands and of parent trees within each stand to be included in cone collections from a region. These standards are probably adequate to prevent loss of much genetic diversity on the regional level.

Unfortunately, they are not always followed. One survey of seed inventories for California showed that seed is frequently unavailable from areas where logging is planned and restocking is mandated. This results as much from the reproductive vagaries of the trees as from any laxity of administration: Douglas-firs produce significant cone crops only every five to seven years, with "lean" years of little or no seed production in-between. When seed is not available for a given area, seed from another zone or another elevation within the zone is often substituted. The amount of acreage actually stocked with such substitutions is hard to document, but it may be as high as 50 percent of all plantings in some areas.

Even when seeds are available from a desired zone and elevation, sampling guidelines designed to maintain genetic diversity may not be met. A 1985 evaluation of Douglas-fir seedlots in the Forest Service Region 5 storage facility revealed that only 16 percent of eighty-seven seedlots contained seed from twenty or more stands as required. The majority of seedlots came from only one to five stands. When such seedlots are used, even the minimum acceptable levels of genetic diversity will not be maintained in the regrowing forest.

There is worry that efforts to fill these gaps in seed inventories can lead to further problems. When cone prices reflect supply and demand, collectors can be tempted to collect seed elsewhere and misrepresent it as coming from an area of low seed availability. Conversations with one Oregon State Department of Forestry geneticist indicated that this is not an idle worry.

Concern over the spotty availability of wild-collected seed, coupled with the potential for producing higher-yielding stock, led the Forest Service to develop its "high-level" tree improvement program. If enough seed could not be collected in the field, why not grow it in orchards, where tree spacing and fertilization could be controlled to maximize seed production? At the same time, the trees planted in the orchard could be carefully selected to give genetic gains in timber production. The goal is to develop seed orchards that will produce enough genetically improved seed to meet the replanting needs of each "breeding zone."

Breeding zones are identified as areas that are reasonably homogeneous in climate, elevation, soils, and other environmental features. "Superior" stands and trees are selected according to Forest Service guidelines emphasizing size and wood production, and cuttings from these trees are grafted onto rootstocks to form a clonal orchard. At the same time, seed is collected from the selected trees and planted at test sites within the breeding zone. Growth of these seedlings is measured and used to determine which of the parent clones will be kept in the orchard for seed production. The process is long and requires large investments in labor and maintenance because trees grafted into the orchard must grow an additional seven to twelve years before they begin to produce significant quantities of seed.

Of course, tree improvement programs are not without their own genetic consequences and concerns. "Domestication selection" is one: Tree breeding programs select not only for families with good growth and yield but also, as a by-product, for clones that can survive grafting without rejection, grow well and quickly in nurseries and orchards, and respond to treatments to encourage early seed production. These are all traits with unknown survival value under natural conditions.

Another by-product of artificial selection is even more difficult to detect. Whereas selective breeding programs intentionally change a small number of traits (such as growth rate) to improve yield, a phenomenon known as "genetic correlation" can change other characteristics along with them. Genetic correlation can be thought of as "group packaging," whereby two traits (say, growth rate and pest resistance, as a hypothetical example) tend to be linked on the same chromosome more often than not. On the average, if you pick a fast-growing tree, it is also more likely to look like lunch to roving caterpillars. For example, G. E. Rehfeldt (1983) found that selection that improved growth rate by about 4 percent increased the number of seedlings susceptible to damage by fall frosts by about 13 percent!

Genetic correlations among all sorts of traits are widespread in living things. More sophisticated selection programs, which "uncouple" the genetically correlated traits and produce gains in growth rate without loss in cold-hardiness, can be developed and are, in fact, in use. But these programs can preserve variation only in traits that someone has identified as worth monitoring. This omits all the other variations in morphology, chemistry, timing, and responses to temperature, light, water, and pests that are coded by the thousands of genes in the Douglas-fir genome.

Replanting with seedlings from field-collected seed or orchard-bred stock raises further concerns. Researchers are uncovering unsuspected reserves of genetic variation and local adaptation on ever finer scales. One particularly detailed study by Robert Campbell (1979) found genetic variations within a single watershed. He calculated that simply transferring seedlings from a north-facing to a south-facing slope at the same elevation could result in as many as 80 percent of the seedlings being poorly adapted to their new site!

Seed transfer guidelines do not address genetic structure or variation on this scale; separation of seeds on any scale finer than the current elevation and seed zones has not been considered logistically or economically feasible. This policy is resulting in the loss of genetic structure within populations and families, replacing it with a wider array of genotypes homogenized over a larger area. Likewise, seed orchards allow trees from widely separated locations within a breeding zone to interbreed, producing seed with combinations of genes from distant sites.

This may seem innocuous enough in a naturally wind-pollinated species like Douglas-fir, but studies have shown that in natural stands, pollen travels on average only about 400 feet, so that matings usually occur only among near neighbors. Most people have heard of some of the negative effects of inbreeding (e.g., the greater chance of rare genetic abnormalities being expressed when too-near relatives marry). Some geneticists now believe that an "optimal level of inbreeding" exists for many plant species, one that weeds out the genetic abnormalities and allows different genes that "work well together" (sometimes called "coadapted gene complexes") to become established in small local breeding units.

In one experiment with alpine plants, Mary Price and Nicholas Wasser (1979) found that the distance between mating individuals that achieved this optimal inbreeding level corresponded closely to the distance that pollen travels in the field. If this finding applies to Douglas-fir as well, then the random matings between "strangers" brought together in seed orchards and clearcuts are destroying a whole level of adaptation that cannot be attributed to the effects of single genes. Breaking up coadapted gene complexes eliminates the balanced action of sets of genes that have come to be associated because they work well together.

The value of genetic variation goes beyond its function as a template for present-day adaptations, though. Much of the genetic variation in Douglas-fir does not have any adaptive value at present, and some of it is maladaptive. For example, there are individuals that begin their annual growth too early for the present Pacific Northwest climate, thus risking damage by frosts every few years, or even every hundred years (Douglas-fir is, after all, a very long-lived organism).

Yet this genetic variation might be essential to the survival of the species under changed environmental conditions. We know that a period of global warming has begun; soon these nonadaptive or maladaptive variants may be favored as the climate rapidly warms. Likewise, other currently nonadaptive variation in plant chemistry or morphology may someday serve to protect the plants carrying these genes from an as-yet-unintroduced exotic pest or disease or a local enemy that has itself evolved a new strategy of attack.

Many examples from agriculture attest to the importance of maintaining genes whose present value is unknown. In addition to the adaptive value of specific genes and coadapted gene complexes, genetic variation has been shown in many cases to have value in and of itself. Agricultural fields planted with a mix of genotypes often suffer less insect damage than do monocultures of a single genetic variety. Similarly, local variation in chemical defensive compounds of several conifers and patterns of defoliation of Douglas-fir by western spruce budworms are all consistent with the conclusion that spatial heterogeneity in forest stands is an effective means of deterring insect outbreaks.

All reforestation options are likely to lower overall genetic diversity by restricting the number of parent trees contributing seeds (and thus genes) to the seedling pool. We have already seen that wild-collected seedlots are often sadly lacking in genetic diversity (when they are not lacking entirely!). The same applies to selective breeding programs: Geneticists calculate that 200 or more unrelated parents should be included in any effort to breed a forest tree, and most current programs target inclusion of 200–300 families. Yet fewer than half the plantations and seed orchards listed in a survey made in the early 1970s met this standard, and many programs today are still short of their targets.

What can be done to keep the vast array of genetic adaptation in Westside forests that we are only just discovering? As with tropical forests, the best way is to preserve intact as much of it as possible, so that researchers can study and catalog its natural workings, gaining insight for future management. This is an important reason for preserving tracts of ancient forest. They are the only places where local genetic structure remains intact.

This discussion has focused primarily on loss of genetic diversity at the within-species level, but replanting often results in a net loss of species as well. If mixes of genotypes within species are important in preventing pest outbreaks and stabilizing community interactions, mixes of species are even more so. Again, most evidence comes from more easily studied agricultural systems, where intercropping of, for example, corn and beans can significantly reduce insect outbreaks and the spread of diseases.

There are other advantages to mixed-species forests. Mixtures including some shade-tolerant species can increase total production by increasing the percentage of the incoming solar radiation that is actually used. And incorporating nitrogen-fixing species can have major benefits: In Washington, a mixed stand of red alder and Douglas-fir had twice the biomass of adjacent Douglas-fir monocultures.

Results such as these confirm that a commitment to genetic and species diversity in forest ecosystems need not be incompatible with higher yields and profitable management. □

VEGETATION MANAGEMENT

What is a weed? A plant whose virtues have not yet been discovered. (Ralph Waldo Emerson, *Fortune of the Republic* [1878])

We often hear from timber industry advocates that forests are a renewable resource and that forest management replaces what it has cut.

How true is this? Among the many ways to answer this question, one concerns ecological competition.

As highly competitive animals, we easily relate to what Tennyson called "nature red in tooth and claw," which evokes images of grizzlies fighting over a gory elk carcass. But oddly enough, in their quiet way, plants live in a more fiercely competitive world. Unlike animals, which can avoid competition by specializing on different food resources, most plants merely need varying *amounts* of the same resources: light, water, and nutrients. Light is often a limiting factor during forest succession. Water can be limiting, especially on upland sites during summer. Nutrients can also be limiting. Whatever resource is limiting, plants must compete to acquire it.

This has important implications for timber operations. Once the ancient forest is cut, the site is burned and "genetically improved" Douglas-firs are planted, and few forest managers let nature take its course because a newly opened site provides an ideal opportunity for other plants at least equally suited to colonizing disturbed habitats. Some are herbaceous annuals that establish quickly but relinquish their hold as they are shaded out. Others are perennial shrubs or trees that cannot replace themselves in their own shade. Douglas-firs have to compete with these species for limiting resources.

"Noncommercial brushfields" now occupy many sites where ancient forests were logged decades ago. Even where Douglas-firs would eventually prevail, it might take a half-century or more. Because other plants can delay Douglas-fir establishment and growth, some foresters consider them "enemies."

Competing species can slow Douglas-fir site capture even on the best sites. But not all sites favor Douglas-firs. On sites too wet, too dry, too cold, or having soils texturally or chemically unsuitable, they are naturally absent. On marginally better sites, Douglas-firs are poor competitors and naturally slow-growing and uncommon. Nonetheless, they have often been planted on unsuitable or marginal sites after ancient forests have been removed.

Facing the sunk cost of seedlings struggling on a site where other plants are superior competitors, what more can a manager do? Manage vegetation, of course. And although many physical methods are available, all have disadvantages in Westside public lands, where slopes and labor costs are both steep. Hence, vegetation management usually means using herbicides.

Herbicides are chemical weapons whose purpose is to kill plants.

Some are fairly selective, eliminating only a few kinds; others are not. Some methods of application confine herbicides to the site; others can distribute them widely. Some herbicides break down quickly into less toxic compounds; others don't. All must be registered with the U.S. Environmental Protection Agency, whose restrictions on their use undoubtedly lessens damage to nontarget organisms. But herbicides can be harmful even when used according to directions. Then again, people don't always follow directions. My father could not have been the only person believing that if a little is good, a lot is better.

Herbicides can markedly alter species composition, structure, and productivity in ecosystems downwind and downstream. But they are a mixed blessing even when they do what they are supposed to do.

What are the effects of eliminating other plants? It depends on what you value. If you are anthropocentric—you see humans as the center of the world—getting rid of plants might, at first, seem to be no problem. Many plants that John Walstad and coauthors (1987) list as principal competing species on Pacific Northwest commercial forestland are not widely appreciated. In fact, some, such as scotch broom, are alien species that do not belong in our forestlands (although hordes of honeybees might disagree) and often displace native species.

But this botanical enemies list also includes the Pacific dogwoods and Pacific rhododendrons whose flowers brighten the forest in spring, the creeping Oregon-grapes, huckleberries, salmonberries, and blackberries that people munch in summer and the vine maples that provide rare splashes of brilliant red and yellow to the deep green autumn forests. It includes the Pacific yews whose bark is a potential anticancer medicine and the cascara buckthorns whose bark provides the active ingredient in most over-the-counter cathartic laxatives. It includes the golden chinquapins on which shiitake mushrooms are cultivated, the California-laurels highly sought by wood carvers and the Oregon white oaks that provide the Westside's best firewood.

The competitors of conifers also include many species that are heavily used by animals, from madrone trees preferred by cavity-nesting red-breasted sapsuckers to alders eaten by caterpillars of pale tiger swallowtail butterflies. As Michael Morrison and Chuck Meslow (1983) found, the presence of these "weed" species in four-to-nine-year-old clearcuts greatly improves the habitat for birds.

Those who cannot see the forest for the trees might dispense with such benefits in a cloud of herbicide. But even when timber production is the only criterion for judging other plants, there is abundant information showing that trees and shrubs considered competitors of conifers can play an important role in Westside forest nutrition.

Among them are alders and *Ceanothus* spp. Throughout the Westside, one or more of these species colonize clearcut and burned sites. Their wood has less commercial value (alders) than Douglas-firs or none (*Ceanothus*). They can, indeed, compete with Douglas-firs for light, water, or perhaps some nutrients. But they are far from useless, for alders and *Ceanothus* spp. are nitrogen fixers par excellence. They supply soil nitrogen that makes Westside forests so productive.

Nitrogen-fixing bacteria need large amounts of chemical energy and low oxygen concentrations. Because such conditions are rare in well-aerated forest soils, free-living bacteria fix little nitrogen. But alders and *Ceanothus* species host bacteria called actinomycetes in special nodules on their roots. The plants provide energy-rich substances and protection from oxygen; the bacteria turn nitrogen gas into ammonia, which the plants then convert to the amino acids and proteins that they need to grow roots, leaves, and wood.

Some of this nitrogen becomes available only when alder and *Ceanothus* die and decompose, the rest when their fine roots die, when they drop their leaves and when grazers eat them and deposit nitrogen-rich scats. Red alders can provide 40–150 kg/ha of nitrogen per year; *Ceanothus* 0–110 kg/ha. Nitrogen fixation at these rates (even for a few years) provides a large pool of nitrogen capital precisely when seedlings need it.

Because nitrogen fixers can compete with conifers, their benefits might be overlooked by foresters impatient to squeeze one more crop of Douglas-firs from the land. But research findings reported by Susan Conard and coauthors (1985) show that conifers grown with *Ceanothus* are inhibited in some cases and benefited in others. *Ceanothus* can outcompete conifers for soil water, but it ameliorates the microclimate, thereby lowering conifer transpiration. It can outcompete them for nutrients or improve their nutrient uptake. It can increase damage from rabbits but decrease aphid infestation. Conard and coauthors warn that *Ceanothus*-conifer systems are extremely complex and urge avoidance of simplistic interpretations. This suggests that reflexive application of herbicides might be unwise even in the short term.

If prodigious nitrogen fixation were not enough, alders have yet another role in Westside forests: They colonize some sites that have been opened when laminated root rot has killed Douglas-firs, hemlocks, firs, or spruces. Alders are nonsusceptible, and they alter soil chemistry in ways that lessen spread of the disease. That is silviculturally valuable, for laminated root rot fungi can prevent successful restocking with Douglas-firs for fifty years until the fungi die out.

Further, Mike Amaranthus and Dave Perry (1989) of Oregon State University suggest that some apparent competitors of Douglas-firs serve the vital role of serving as hosts to the same mycorrhizal fungi that colonize Douglas-fir roots, thereby maintaining mycorrhizal populations essential to Douglas-fir establishment in the years after logging.

So, even with all the advantages that modern genetics can provide, the "superior" Douglas-firs that are planted often grow slowly or succumb when faced with competitors shaped by natural selection. Despite all the manipulations used to stack the deck, planted trees can fail to maintain adequate stocking levels in the face of natural competitors. Perhaps nature is telling us something. But some of these species that might lessen economic returns in the short term perform important, even indispensable, services for timber growers, increasing economic returns in the long term. It is mystifying that foresters who seek to turn forests into timber farms have ignored an elementary lesson of successful farming: that crop rotation and fallowing fields renew the land. These "weeds" could well be essential to sustainable timber production.

But timber production is not the only value of forestland. Everywhere we look, we see that other plants have economic, ecological, and aesthetic values. It is easy to forget that Douglas-firs coexisted with them and succeeded them for eons, covering great acreages with superb timber trees without any help.

So, should other plants be eliminated at all costs or maintained for the roles we are only coming to appreciate? The answer depends on our values.

PEST AND DISEASE MANAGEMENT

Every sweet has its sour, every evil its good.
(Ralph Waldo Emerson, *Essays: First Series* [1841])

Within a day's drive of Washington, D.C., gypsy moth larvae are munching through large areas of eastern forest. Balsam woolly adelgids are killing Fraser firs already stressed by air pollution. Dutch elm disease fungi are choking American elms. And the magnificent American chestnut, once the most important tree in many eastern forests, is virtually gone, strangled by chestnut blight fungi.

Nobody likes to see these trees dying. But all these epidemics are alien. Humans brought them to eastern forests. In contrast, most pests and pathogens of Westside conifers are natives.

This distinction might seem academic to a forest manager losing a fine stand of Douglas-firs to native laminated root rot, but it is not. Trees that have never encountered alien insects and pathogens and have therefore not evolved specific defenses against them cannot deter some aliens from becoming devastating plagues. In contrast, trees have had the time to evolve deterrents against native insects and pathogens, so that it is mainly those weakened by competition, drought, fire, or logging damage that succumb. After coevolving for millennia, native insects and pathogens would not threaten trees in natural Westside forests as aliens are now doing in the East, so long as nothing tilts the balance in their favor.

There is ample scientific evidence that some "enemies" of trees are important for maintaining biological diversity and that knee-jerk reactions to control insects and fungi can damage the health of Westside forests. Many of us were taught that insects and fungi, like predators and fires, are "bad." But, as often the case in living systems, the truth is more complex.

Perhaps the most notorious "enemy" of Westside trees is the native fungus *Phellinus wierii*, which causes laminated root rot. It diminishes Douglas-fir production by perhaps 5–10 percent. How can such a pathogen be anything but bad?

Scientists will probably learn many answers as we begin to look at the intricacies of coevolutionary relationships between the fungus and its hosts. But a fascinating one concerns western redcedars.

Western redcedars should be rare in most places because they are inferior competitors of other major Westside trees. Although tolerant of shading, they are slower-growing and less prolific than the even more tolerant western hemlocks. They have much thinner bark than Douglas-firs, making them vulnerable even to cool fires. And they cannot withstand cold as well as mountain hemlocks, firs, and Sitka spruce, which excludes them from high mountains and colder areas north of the border. They do tolerate waterlogged soils better than these others and can inhabit swamps and riparian zones from which the others are excluded, so they have a refuge from competition. But redcedars also occur on hillsides with no standing water, where their competitors would seem much better adapted. How do they do it?

An answer comes from an unexpected corner of the Pacific Northwest: the sea. More than two decades ago, the University of Washington's Bob Paine (1966) studied the role of the predatory ochre starfish in Puget Sound. In the rocky intertidal zone, blue mussels overgrow and smother acorn barnacles, gooseneck barnacles, limpets, and chitons where the starfish is absent, yet all occur in the starfishes' presence. Paine's experiments demonstrated that the starfish, by preferentially removing the competitively superior mussels, allows its other prey species to live there. Thus, the predator *increases* species diversity. Indeed, Paine later dubbed the ochre starfish a "keystone species," one whose activities are crucial in determining the composition of its community far out of proportion to its abundance.

The villainous pathogenic *Phellinus wierii* is no less a keystone species than the voracious predatory ochre starfish. Douglas-firs, hemlocks, firs, and spruces are susceptible to laminated root rot. Redcedars (and incense cedars, Port Orford cedars, Alaska-cedars, and various broadleaves) are immune. By killing susceptible trees and preventing their reestablishment, laminated root rot can allow redcedar seedlings to establish where they would otherwise be outcompeted.

The benefits of *Phellinus wierii* extend even to the very species it infects. By eliminating some competitive dominants and making room for nonsusceptible species, *P. wierii* creates a natural mosaic community pattern that helps to limit its spread; it is self-limiting. And the pattern it induces also limits the spread of other pests.

Insects and pathogens are also natural agents of thinning: They cull the weak. Healthy trees have defenses against attack, but most defenses require energy. Stressed trees have less energy to devote to repelling invaders such as bark beetles. If enough invaders breach the defenses of a weakened tree, it dies. In doing so, it leaves more resources for nearby trees, accelerating their growth and reproduction, and it becomes a snag or log, creating structures crucial for many wildlife species. What seems "bad" at first glance can have important benefits when we look more carefully.

Westside trees are hosts to some native insects that become epidemic in the Eastside, including Douglas-fir tussock moths and western spruce budworms. But healthy trees in natural Westside forests seem amazingly resistant to epidemics of native pests and diseases. Unfortunately, like eastern forests, they are less resistant to aliens.

Balsam woolly adelgids, aphidlike sucking insects introduced from Europe early in this century, are epidemic not only on some firs in the Appalachians but also in the Northwest. Within five years of initial attack, they can kill up to 90 percent of subalpine firs in some stands. They also plague Pacific silver and grand firs. Whether or not management can help these species is uncertain. Their future could hinge on whether they have enough genetic diversity to resist this unprecedented onslaught.

Introduced white pine blister rust fungi eliminated perhaps 95 percent of western white pines earlier in this century. Their loss has harmed not only the timber industry but also species (such as brown creepers) that use white pine snags as nesting sites and huge numbers of wild currants and gooseberries (the fungus's alternate hosts) destroyed in attempts to control the disease. In some areas of the West, white pine blister rust has also harmed whitebark pines, whose cones provide essential forage for grizzly bears.

Both timber producers and native species have been hurt by these introduced diseases. Another introduced epidemic of Port Orford cedars is so closely tied to road-building and logging practices that it is best examined in light of the cumulative effects of timber operations.

FIRE SUPPRESSION

According to its fuel so doth a fire burn.
(Ben Sira, *Book of Wisdom* [ca. 190 B.C.E.])

In the religion of ancient Greece, the other gods condemned Prometheus to eternal punishment for giving godlike power to humankind: the ability to control fire. Now, in a technologically advanced society that routinely uses fire to forge metals, generate electricity, and propel automobiles, the awesome power of fire can still grip our consciousness. For fires sometimes happen where and when we do not intend, and, faced with their power, we feel compelled to try to reassert control over nature.

The Forest Service has been very successful in getting this message across. Even in cities far from national forests, Smoky the Bear's message, "This was once a forest," reaches people in schools and subways. And many other shapers of our beliefs have reenforced it. Few children remain untouched when they see Bambi and Thumper fleeing the devastating wildfire. Our mythology teaches that fire is good when we control it but bad when we do not.

But when ecologists look at fire, we are struck by its central role in many forest ecosystems. We see that some species are vulnerable to fires, others are resistant, and still others depend on fires to survive or reproduce.

Major fires require (1) lots of fuel, (2) circumstances that favor their propagation, and (3) a source of ignition. Westside ancient forests are unusual in all three.

1. *Fuels.* The enormous biomass of ancient forests provides some of the world's highest fuel loadings, creating the potential for great conflagrations. However, their fuel size distribution is not ideal for propagating conflagrations under most circumstances.

Fine fuels—needles, twigs, shrubs, litter, and duff—ignite faster than large branches, standing trees, and fallen logs for two reasons. First, woody fuels ignite faster when their water content is lower because heat is needed to drive off water. Because fine fuels have high surface area per unit volume, wind, high temperatures, and low humidities dry them much faster than coarse fuels.

Second, fine fuels require much less heat to combust. That is why a match can ignite a dry piece of newspaper but not an equally dry three-inch-diameter log. In the presence of enough fine fuels, wildfires can spread very quickly.

But when a fire spreads too quickly, it does not have enough time to heat larger fuels and set them afire. Further, fine fuels release relatively little heat, so huge amounts are necessary to produce the sustained heating needed to ignite green trees 3, 6, even 9 feet in diameter or large downed logs that typically have an internal water content of 250, even 350 percent of their dry weight. As a result, fires consume a much larger proportion of fine fuels than coarse ones. Only rarely are giant trees and logs consumed. That is why many remain when fires pass through Westside forests.

2. *Circumstances that favor propagation.* The Westside has a distinct fire season; few fires can sustain themselves in the cool, wet months. Because summers are often foggy and at least a few percent of the prodigious Westside rain falls in summer, most summer droughts cannot dry the large fuels that support major conflagrations. Under exceptional drought conditions, however, especially when hot, dry, easterly foehn winds pour down through the Cascades passes, conditions are set

for big fires. All that is needed is an ignition source. But here too the Westside is unusual.

3. *Ignition sources.* Volcanic eruptions can ignite fires. The impact of an asteroid might have set fires worldwide 65 million years ago. But such ignitions are very rare; nearly all natural fires are started by lightning.

The Westside averages fewer than five thunderstorm days a year, compared with 80–100 in parts of south Florida. So, despite the huge fuel loadings, lightning strikes coincide with conditions favoring propagation of hot fires so seldom that fires large and hot enough to kill entire stands are very unusual. Nevertheless, what is rare on human timescales becomes all but inevitable on the millennial timescale of giant conifers.

In the wettest, foggiest forests—redcedar swamps and coastal Sitka spruce/western hemlock forests—fire might well be so rare that it is inconsequential. In slightly drier Douglas-fir/western hemlock forests in northern Washington, a given spot probably has catastrophic fires no more than a few times a millennium; few stands exceed 750 years. Fires are more frequent farther southward in the Cascades and are especially frequent in the warmer, drier areas of southwest Oregon and northwest California. But because frequent fires tend to eliminate the most readily burned fuels, they do not burn as hot as infrequent fires; few are hot enough to kill whole stands. An increasing fraction of the fires farther south are "underburns" or patchy "partial burns" rather than big stand replacement fires.

Still, biomass in Westside forests is so high that conflagrations do occur. Certain broad age classes of trees are rare, but others are widespread, suggesting that stand replacement fires—which destroy extensive areas of canopy and allow shade-intolerant species to germinate and grow—occur at discrete and infrequent intervals. Today's pattern of ancient forests probably results mainly from very rare (once in centuries) fires of astounding magnitude. For example, the many sites in the southern Washington and Oregon Cascades where dominant Douglas-firs are around 450 years old probably resulted from one or a series of fires 450–550 years ago.

Major fires initiate episodes of death and regeneration. Minor fires favor some species and age classes over others. Fire is as much a part of Westside forests as Douglas-firs, cutthroat trout, and spotted owls, species that have evolved in ecosystems shaped by fire.

Of course, humans have altered Westside fire regimes. Native Americans burned some drier lowland areas to encourage preferred food plants such as bracken ferns and camas, but Euro-Americans have altered the fire regime far more. We have altered the fuel distribution. We are changing the climate. We have increased the number of ignition sources. After building homes and businesses in forests and deciding that forests are our economic dominion, we have felt compelled to protect these interests. And so, people have suppressed fires as a matter of course.

Fire suppression has many ecological effects. It changes species composition by allowing the survival of organisms that would have been killed while it diminishes opportunities for species that fare better after fire. It reduces tree death rates, thereby reducing the number of new light gaps, snags, and logs, thereby reducing diversity of shrubs, herbs, and cavity nesters. And fire suppression can have another well-known but paradoxical effect: increasing the risk of catastrophic fire.

This effect seems weak in moist tropical evergreen and temperate deciduous forests, where the decomposition of leaves and wood is quick. But decomposition in coniferous forests is slow, especially in cooler northern, and montane forests, so fuels accumulate, which can increase the intensity and hasten the spread of any fire that gets started.

Because effective fire suppression in the Northwest began only in the early 1900s, its effects on fire intensity are less visible in forests that naturally burn once in centuries than in those that burn more often, such as the Ponderosa pine forests east of the Cascades.

Ponderosa pines are not climax trees, but large ones have thick bark that protects their cambium layer, and they tolerate fires well. As Jim Agee (1981) of the National Park Service explains, in the absence of fires, they form open, parklike stands that are invaded by grasses, shrubs, and shade-tolerant (but less fire-tolerant) trees such as grand firs. Under natural regimes, these are eliminated by fires that sweep through the stands at intervals averaging five to twenty years, leaving most of the big pines undamaged.

But if fire suppression allows undergrowth to accumulate, the higher fuel loading near the ground allows fires to burn hotter. Further, the shade-tolerant firs show less self-pruning than the early successional pines. The branches that firs retain form a "fire ladder" that conveys surface fires into the crowns, where they spread to ponderosa pines. When such fires cannot be controlled, they can eliminate grasses, shrubs, firs, pines, and so on—all the vegetation. Thus, fire suppression in ponderosa pine forests is like applying insecticides to crops or smoking cigarettes: The longer you do it, the harder it is to stop.

The Westside forests with fire regimes most like those of Eastside ponderosa pine forests are the mixed conifer forests (which can include ponderosa pines) and mixed evergreen forests of northwest California and southwest Oregon. Fire frequencies naturally decrease to the north in the Coast Ranges and Cascades with decreasing summer temperatures and shorter droughts. But charring on the trunks of living trees shows that even the wet Douglas-fir/western hemlock forests in the Olympics can have fires that reduce fuel loadings but are not hot enough to kill the dominant Douglas-firs. This pattern suggests that the risk that fire suppression will cause catastrophic fire is probably greatest in northwest California and southwest Oregon, decreasing to the north. Franklin and Hemstrom (1981) feel that fuel buildup due to fire suppression does not affect fire regimes at Mt. Rainier. Presumably this is no less true in still cooler and wetter forests.

Finally, fire suppression has another effect that can increase the severity of fires: By eliminating the natural gaps caused when minor fires kill some dominant trees, fire suppression produces a continuous canopy that can promote the spread of crown fires.

Cumulative Effects of Timber Operations

SPECIES DIVERSITY

The worst thing that can happen—will happen—is not energy depletion, economic collapse, limited nuclear war, or conquest by a totalitarian government. As terrible as these catastrophes would be for us, they can be repaired within a few generations. The one process ongoing in the 1980s that will take millions of years to correct is the loss of genetic and species diversity by the destruction of natural habitats. This is the folly our descendants are least likely to forgive us.
(Edward O. Wilson, "Resolutions for the 80s," *Harvard Magazine* [1980])

To the untrained eye, the effects of timber operations can be dramatic or subtle. A person might see moonscape clearcuts or thick masses of young trees, spaghetti roads or landscape quilts of green and brown. But aesthetics are imperfect guides for gauging the ecological effects of timber operations. Many effects are difficult to see because they occur on scales that our senses are not designed to perceive. Some are visible only from the air; some are apparent only on time spans longer than a human life. But standing back, one thing becomes clear: The effects of timber operations are hardly trivial.

Looking at phases of timber operations one by one highlights their effects. But these phases don't happen in isolation; they happen in sequence, so at any given site, effects are cumulative. Further, they are changing the whole landscape cumulatively during an interval that is long by our standards—a life span or two—but only a small fraction of a Douglas-fir's life span. It is important to consider timber operations as a whole because they are happening as a whole.

It is necessary to start with a simplifying assumption: that only timber operations are causing ecological change in Westside forests. Later I will discuss how they interact with major external factors. Logging has cumulative effects on virtually every aspect of the land, including ecology, hydrology, and geomorphology. This section examines how logging cumulatively influences three key aspects of forest ecosystems: species composition, susceptibility to insects, pathogens, and fire, and the nitrogen budget.

Some effects of timber operations are invisible; until they become acute, detecting them requires expensive scientific equipment. Some are easy to miss unless the timing is right. But if you know what to look for, the effects on species diversity are unmistakable.

In theory, forest management can have many purposes, but the main purpose of most Westside forest management is timber production. As Gordon Robinson (1988) points out, forests managed for sustainable timber production *can* strongly resemble natural forests. But the practices used in the Westside affect species composition so much that it is difficult to mistake intensively managed sites for natural forests.

Biological diversity is not just a numbers game. If it were, then ancient forests would be less important than they are because they do not usually have the highest species diversity during natural forest succession. A several year-old clearcut can have *more* species. But by and large, these are disturbance-tolerant species (such as dandelions and brown-headed cowbirds) that are abundant and widespread because humans have increased disturbance frequency and intensity so much. This does not apply to Pacific yews and spotted owls, which are growing rarer as ancient forests are destroyed, fragmented, and simplified.

Maintaining species diversity means more than maximizing numbers of species. It means preventing naturally occurring species from becoming uncommon or extinct. Anyone seriously interested in diversity would have no difficulty selecting those to be saved when given the choice between a diverse assemblage including thousands of black rats, common pigeons, and German cockroaches or a breeding pair of Siberian tigers. The world does not lack for German cockroaches.

Recent studies by Raphael, alone (1988) and with coauthors (1988), use abundance of land vertebrates in different successional stages to project how logging might affect land vertebrates in northwestern California. They contrast prelogging forests, today's forests, those most likely in the future, and a worst case (Table 6.2). They estimate that clearcuts are now as abundant as they will ever be and that closed-canopy forests will dominate future landscapes about as much as mature and old-growth forest dominated the natural landscape.

Table 6.2

PAST, PRESENT, AND FUTURE DISTRIBUTION OF STAND AGES OF
DOUGLAS-FIR FOREST IN NORTHWEST CALIFORNIA

When	Mature/Old-Growth (>150 years)	Pole/Sawtimber (20–150 years)	Brush/Sapling (<20 years)
Prelogging	74	13	13
Present	37	18	45
Likely future	11	71	18
Worst case	0	78	22

SOURCES: Raphael (1988); Raphael et al. (1988).

This mix of successional stages suggests that some species (such as brown-headed cowbirds, creeping voles, and western skinks) will become far *more* abundant as ancient forests disappear. But it also suggests that other species (including warbling vireos, northern flying squirrels, and Del Norte salamanders) will become much rarer.

Even closely related species will have markedly different fates. Dusky flycatchers and house wrens are likely to increase by 216 percent and 41 percent, respectively. In contrast, Hammond's flycatchers and winter wrens could decrease by 43 percent and 68 percent, respectively, or more in the worst-case scenario. These projections could well be too optimistic because stands these researchers studied had significant numbers of snags and downed logs. Unless future management schemes deliberately create enough usable dead trees throughout the rotation, the many species that need them will undoubtedly suffer even sharper declines.

There is no reason that one flycatcher or wren species is *inherently* more valuable than another. But ancient forest species are rarer and at greater risk of disappearing. That is why they need special attention in management schemes.

Years ago, we often heard that old-growth forests are cellulose cemeteries or biological deserts (those who used the latter term must have been unfamiliar with deserts, which are seldom biological deserts!) and that managed forestlands are ideal for wildlife.

Scientific understanding has come a long way since then. In recent years, wildlife biologists have realized that these statements devalue any living things that hunters do not enjoy shooting. As fish and game departments have become wildlife departments, nongame animals and plants have been recognized as important components of our biological wealth. There is increasing recognition that good land management values *all* living things, whether or not they are shootable, fishable, or usable as timber.

Knowledge has evolved along with our values. Research in the last two decades has shown that ancient forests are rich in species that take advantage of their special resources, spatial structures, and microclimates. Ancient forests are not only diverse; they are stable, enduring storms, fires, insect outbreaks, and diseases for centuries with little basic structural or functional change.

In a natural landscape, the high-diversity stage that begins in the years after disturbance is short-lived, perhaps two decades. The low-diversity, closed-canopy stage lasts perhaps two centuries. The high-diversity ancient forest stage lasts at least several centuries, sometimes a millennium. Most of a natural landscape is in ancient forest at any given time.

In contrast, intensive forestry practices suppress species-rich early successional stages by spraying herbicides, planting densely, and fertilizing to encourage canopy closure. Trees are cut at 110, eighty, even forty years, so that species-rich old-growth never develops. Hence, at any given time, most of a managed landscape is in the closed-canopy stage, in which diversity is lowest.

The National Forest Management Act requires the Forest Service to maintain species diversity. But any objective look at the evidence shows that current intensive timber management practices have precisely the opposite effect.

PEST AND FIRE REGIMES

Koyaanisqatsi (*Hopi*): *Nature out of balance* (*noun*): . . . *a way of life so crazy it calls for a new way of living . . . applicable to the kind of misuse of technology that creates ecological or human catastrophes.*
(Howard Rheingold, *They Have a Word for It* [1988])

The recent flowering of landscape ecology has stimulated scientists and managers to look in new ways at Westside landscapes in transition. Two changes already well under way are the spread of logging roads into the last redoubts of virgin forest and the replacement of ancient forests by simplified, closed-canopy tree plantations. As Dave Perry (1988) of Oregon State University explains in a thought-provoking article, these changes will likely increase the spread of pests and wildfires across Westside landscapes.

Logging roads carry ignition sources (recreationists with their cigarettes and campfires and loggers with their slash burns and sparking chainsaws), the egg cases of insect pests such as gypsy moths, the seeds of noxious weeds such as tansy ragwort, and the spores of pathogenic fungi. One such pathogen, *Phytophthora lateralis*, causes Port Orford cedar root rot, a disease that is rapidly eliminating a tree endemic to scattered sites in southwest Oregon and northwest California. A look at how roads spread this disease hints of their importance in dispersing all sorts of organisms that could not penetrate the forest primeval.

Port Orford cedars are highly sought because their lumber resembles that of hinoki, a now scarce species that Japanese hold in religious veneration. Virtually the whole cut is exported to Japan. They have a ready market even when the timber industry is depressed and bring much higher prices (averaging $2,166 per 1,000 board feet from one district in 1981) than more common conifers, such as Douglas-firs. Indeed, they are so desirable that the timber industry has overcut them throughout this century and virtually eliminated larger trees.

Although little is known about Port Orford cedar's genetic diversity, breeders of ornamentals have developed at least 220 cultivars with exceptionally varied growth forms and foliage colors, suggesting that genetic variation is considerable. Port Orford cedars regenerate well in clearcuts, appear to be less sensitive than other conifers to a wide range of temperatures and cold, wet, or poor soils, are the only commercially salable trees that can grow on some sites, resist fires well (old trees have very thick bark), and are unusually free of native pathogens and insects. These qualities would make them extraordinarily valuable for the timber industry in the future, except for Port Orford cedar root rot.

P. lateralis is an apparently introduced aquatic fungus that always kills its host; no resistant Port Orford cedars have been found. It was first discovered in Port Orford cedars in nurseries, and it began killing trees in their natural range in about 1950, a time when the postwar housing boom had accelerated construction and logging. It spreads through root-to-root contact or via spores borne in water or in soil. In the wet season, it is also spread by elk, cattle, and machinery, including road maintenance and logging equipment, trucks, and off-road vehicles. It can move downhill rapidly via streams. Its uphill movement is slow, but the expansion of logging roads has greatly sped dispersal among the scattered upland Port Orford cedar enclaves.

The fungus does not respond to any treatment; the only current management strategy is to prevent its spread. Donald Zobel and coauthors (1985) recommend minimizing the dispersal of spores during road construction, road maintenance, and salvage logging and keeping tree plantations away from roads. Specific recommendations include carrying out timber operations only during the summer dry season, closing logging roads, and using pressure cleaning to remove mud from vehicles working in infested areas. Some of these steps require unprecedented cooperation among all who use logging roads, from timber companies to off-road vehicle users. Without the roads the disease would, at most, spread uphill among the scattered sites very slowly. But with logging roads expanding into the last parts of its range, Port Orford cedar, which was proposed for listing under the Endangered Species Act, is probably doomed commercially and possibly headed for extinction in the wild.

Roads are not the only way that logging spreads threats. Current forest management—particularly clearcutting and planting—is replacing biologically diverse natural forests with genetically less diverse, even-aged, closed-canopy monocultures, which can favor pests and diseases over their hosts.

Monocultures create ideal conditions for the uncontrolled spread of pests and diseases, as shown by Cornell University's Dick Root (1973). He found that collard plants grown among other species suffered much less insect damage per plant than those grown in monocultures. The reason is related to the incredible reproductive potential of most pests.

Finding a suitable host is probably the greatest problem an insect pest or pathogen faces, and most produce many propagules (dispersing stages, such as crawling larvae, flying adults, or spores borne by the winds) to overcome it. Where the diversity of vegetation is high, many dispersing pests land where they cannot feed or reproduce. Further, the structural complexity of more diverse communities provides habitat for many more predators, including birds, spiders, and predatory insects. Together, these factors cause high pest mortality.

But in monocultures, without barriers to dispersal, insects and pathogens find unlimited resources in all directions. With sharply lower mortality, pest populations grow without constraint until food becomes limiting. By that time, a species that might have been harmless in a biologically diverse landscape can be severely damaging.

This is as true in crops of conifers as in collards or apples. For example, timber management increases two fungal diseases, Annosus root disease and the more worrisome black-stain root disease. As Ev Hansen and coauthors (1988) point out, black-stain root disease is most common near roads in young, overstocked, pure Douglas-fir stands that have been disturbed by heavy equipment and thinning. Although its incidence is still low, black-stain root disease is increasing at epidemic rates as its insect vectors spread among ever-expanding tree plantations.

Old-growth forests provide opportunities for far more species of arthropod predators than even-age tree plantations (Table 6.3), which helps to limit populations of insect pests. Indeed, the diversity of all three arthropod functional groups (guilds) is far higher in an old-growth stand than in a regenerating tree plantation. But the biomass patterns of

Table 6.3

BIOMASS AND SPECIES RICHNESS IN CANOPIES OF DOUGLAS-FIRS IN OLD-GROWTH (400-YR.-OLD) FORESTS AND REGENERATING (10-YR.-OLD) MONOCULTURES IN WESTERN OREGON

	Old-Growth Forest		*Regenerating Monoculture*	
Feeding Groups	BIOMASS (GR/HA)	NUMBER OF SPECIES	BIOMASS (GR/HA)	NUMBER OF SPECIES
Herbivores	190	8	370	3
Predators	160	40	50	10
Others	30	18	0	2
Total	380	66	420	15

SOURCES: Schowalter (1989); Schowalter and Means (1988).

the three feeding guilds (groups of species having similar ecological functions) differ dramatically. The biomass of predators in old-growth is twice that in the monoculture, while the biomass of herbivores— potential pests—is five times greater in the monoculture.

Tim Schowalter and Joseph Means explain that finding a suitable host is difficult for short-lived insects or spores in complex ecosystems and that old-growth forests, with their diversity of tree species, age classes, and predators, "are less conducive to pest outbreaks than are the simplified forests created through current harvest and regeneration practices" (1988, p. 342).

Furthermore, as Seri Rudolph explained earlier, some conifer species have exceptional genetic diversity, which confers diverse defenses against pests and pathogens. For example, in ponderosa pines, adjacent individuals can show very different levels of insect infestation. Genetic diversity deters pest outbreaks by offering diverse suites of defenses.

Monocultures planted with stock from selected seeds or clones offer less diverse defenses than natural stands. Breeders select almost exclusively for fast growth and good form, traits that maximize profits *under ideal conditions*. But selecting *for* any trait selects *against* all other traits with which it is not genetically linked. Hence, the widespread even-aged, genetically simplified monocultures of the near future will be vulnerable to outbreaks.

Fire resistance is also critical. Young trees are smaller and thus require less heat to ignite. They have thin bark that offers little protection to underlying living tissues, including the meristems essential to radial growth. Although dense young stands (such as tree plantations) lack ground cover, they have large loadings of fine branches near ground level that can carry fire into the crowns. Their crowns are closer to the ground, so shorter flames can ignite them. And the continuous canopy poses no barrier to the spread of fire. As Franklin and Hemstrom (1981) point out, they are vulnerable until they reach 75–100 years.

As the stand ages, the amount of heat needed to ignite a tree increases, the bark thickens, and fungi help trees to "self-prune" lower branches. As the fuel succession approaches its low point and canopies rise out of the reach of flames, the stand becomes less vulnerable. Mature forest is the least flammable successional stage.

Ancient forests have characteristics that both promote and dampen fires. Like mature trees, old giants usually have their first branches high up, which inhibits the spread of surface fires into the canopy, and their bark is even thicker. But clumps of shrubs and shade-tolerant understory trees create a fuel ladder that can convey fire into the crowns of old-growth trees.

However, the discontinuous canopy of an ancient forest can inhibit the transmission of crown fires. So can the diversity of tree species, at least in mixed evergreen forests of southwest Oregon and northwest California. In 1987, when fires burned 100,000 acres of Siskiyou National Forest, Perry (1988) observed that tree plantations suffered far more damage than old-growth forests, where madrone trees in the understory seemed to inhibit the spread of fire.

Using a fuel model for natural forests, Jim Agee and Mark Huff (1987) suggest that fires spread fastest in the youngest stands, slowest in mature (over 100-year-old) stands and at intermediate rates in old-growth. Since most tree plantations will be cut as soon as they reach commercial maturity, a very large fraction of the landscape will be occupied by young stands ideal for spreading fires. By increasing the risk of pest outbreaks and wildfire, converting old-growth forests to tree plantations threatens not only biological diversity but the health of the timber industry and the stability of Westside landscapes.

NITROGEN BUDGETS

Whatever you have, spend less.
(Samuel Johnson, in James Boswell, *The Life of Samuel Johnson* [1791])

As explained earlier, nitrogen budgets in forests, like any budgets, are affected by both income and outflow. Westside forest ecosystems maintain their high productivity both by maximizing inputs and minimizing outputs during succession. In contrast, intensive timber operations diminish nitrogen inputs while increasing outputs, thereby depleting the ecosystems' nitrogen capital.

To see how, we can compare balance sheets of a natural forest and a tree plantation. Ecosystems get some of their usable nitrogen as fallout from the atmosphere. Areas downwind of concentrations of power plants or automobiles (southern California, the eastern United States, Europe, eastern China) get very high annual nitrogen inputs. This is both a blessing and a curse: It provides a vital plant nutrient but can damage ecosystems since it arrives mainly as nitric acid.

However, prevailing air masses from the Pacific are clean. Away from Westside cities and industrial areas, inputs average only 1–2 kg/ha per year. They are small compared with amounts that trees need.

Ancient forests have two ways of maintaining high productivity in the face of scarce atmospheric nitrogen inputs. They have some exceptionally effective N-fixing species, and they are highly conservative: They hang on tightly to whatever N they get.

If a disturbance is severe enough to destroy an old-growth forest canopy (especially if it is a fire), red alders or *Ceanothus* spp. will probably colonize the site. These N fixers contribute large amounts of N for a few decades, although eventually their contribution declines. At a very conservative annual rate of 40 kg/ha, they can add a minimum of 1,000 kg/ha to the soil before being overtopped by conifers, perhaps several times as much. Another 1 kg/ha is fixed each year in downed logs, and a like amount comes from the atmosphere. And, as Phillip Sollins and his coauthors noted, "In old-growth stands, cyanophycophilous lichens fix N at rates approaching 10 kg ha^{-1} year^{-1} ... but these lichens are absent during the first 100–150 years of stand development" (1987, p. 1594). Again, being conservative, we can say the yearly N contribution from canopy lichens averages 5 kg/ha after age 150. A small sum is added by free-living N fixers in the duff and soil, say 1 kg/ha per year, as Warwick Silvester and coauthors (1982) found.

Together, these diverse sources provide over 4,000 kg/ha of N by the time a stand reaches 500 years old; 500 kg each from the air, N fixation in logs and in duff; 1,000 kg from early successional N-fixing plants and 1,750 kg from N-fixing lichens. Some N is stored in the soil; the rest is in biomass.

Of course, all systems have leaks, and ancient forests are no exception. About 1 kg/ha of N leaves in streamflow each year, so, assuming that losses are no greater in younger stages, about 500 kg/ha is lost during the life of a representative stand. Even if the stand is devastated by a fire that volatilizes another 1,500 kg/ha, the ecosystem has accumulated a lot of N capital, perhaps 2,000 kg/ha. Not surprisingly, the soils of some Westside forests, especially in coastal fog belts (where fires are very rare) can have huge amounts: 25,000 or more kg/ha, among the world's highest accumulations of soil N.

The N budgets of tree plantations are rather different. First, the roads that take up perhaps 8 percent of the area receive virtually no N from early successional plants, late successional lichens or free-living bacteria in logs and soil. They contribute only what falls on them as rain and dust and washes off into the forest. Thus, roads reduce N inputs by close to 8 percent.

Compared with losses from later phases of timber operations, that's small potatoes. The need to maximize profits means that an ever greater number of disturbed areas (burns, blowdowns, or, increasingly, clearcuts) outside national parks and wilderness areas are subjected to vegetation management to suppress early successional plants such as red alders and *Ceanothus*. Although vegetation management might speed conifer establishment, it robs trees of their nitrogen "birthright" just when they need large amounts to make new foliage and roots.

This might not be so bad if other sources continued to contribute, but tidy foresters remove big cull logs during site preparation, eliminating one source. Further, a standard rotation is too short for canopy lichens to contribute, eliminating another far more important one. Assuming that free-living soil N fixers are not affected, only they and atmospheric inputs replenish N capital in a tree plantation. The total input per eighty-year rotation will be perhaps 160 kg/ha.

Assuming a tree plantation loses N at the same rate as old-growth, during that interval, 80 kg/ha will leave in streamflow. Thus, by the time it is ready to be logged, instead of at least 3,500 kg/ha in a 500-year natural "rotation," only 80 kg/ha have accumulated.

But that is not the end of the story. Rather than just failing to build N capital, timber operations put forests deeply into debt. Although logs are low in N concentration, their biomass is very large. Sollins and McCorison (1981) point out that removing them can take out nearly 400 kg/ha. Further, to prepare the site for the next rotation, slash (which includes N-rich foliage and twigs) and duff are burned. Susan Little and Janet Ohmann (1988) of the Forest Service found that N loss from burning the duff varies so much that they declined to provide an average, but the median change was a loss of about 290 kg/ha from the duff alone (i.e., not including the amount lost from the slash). Then there is N loss in streamwater following removal and burning; Fredriksen and Harr (1981) say that it can be as high as 200 kg/ha and probably ranges from 50 to 140 kg/ha, although Sollins and McCorison say that losses are much lower.

Actual numbers vary depending on how intensively forests are managed. But no matter how you cut it, current timber management practices *decrease* forest ecosystem N capital by at least many hundreds, and probably over 1,000 kg/ha per rotation, while natural forest succession produces net *increases* several times greater. Of course, to replace the N that natural processes provide, foresters could add tons of N fertilizer per hectare. And some industrial owners of intensively managed tree plantations do add at least some N, perhaps a few hundred

Hydrological station in H.J. Andrews Experimental Forest, Willamette National Forest (Oregon). *These stations are used to analyze the water leaving both pristine and clearcut watersheds. Because ancient forests release very little nitrogen into streams, the water they produce is some of the purest in the world. After clearcutting, nitrogen loss increases. Nonetheless, it is only 0.5 percent of the nitrogen lost when the logs are removed. Intensive timber operations sharply decrease the gain and increase the loss of this vital nutrient from the forest.*

pounds per hectare each rotation. But this does not compensate for what is lost.

There are some problems with fertilizing forests artificially. For one, it is unclear whether fertilization itself decreases N fixation. F. J. Stevenson (1986) says that N fixation is sharply reduced in the presence of readily available combined N. Conard and coauthors (1985) cite a finding that high soil N suppresses N fixation in both legumes and other families of plants. In contrast, Dan Binkley (1986) says that high soil N might inhibit N fixation by legumes, but that N fixers that depend on actinomycetes (alders and *Ceanothus* spp.) can fix large amounts even in N-rich soils. Even if Binkley is correct, the point is moot when these N-fixing plants are killed with herbicides.

Further, fertilizing ecosystems with nitrogen leads to elevated N levels in runoff, which increases growth of freshwater algae. In very small amounts, this is probably beneficial because it increases stream productivity. But in larger amounts, it causes over enrichment (eutrophication). People who have studied the water quality of streams that drain croplands or feedlots will point out that eutrophication is anything but beneficial.

However, many tree plantations that replace ancient forests—especially on public lands—will *not* be fertilized. Fertilizing is already expensive. And since manufacturing N fertilizers requires large amounts of natural gas, prices are likely to rise sharply as competition for our shrinking natural gas resources increases in a world that is hungry for cleaner, less carbon-intensive fuels. Loggers will understandably choose not to replace what they do not own. Public land managers are far more likely to drain invisible soil N reserves than spend precious funds on fertilizers, so that some other ranger might meet his or her timber quota eighty years from now.

Even this cursory nitrogen budget suggests that intensive timber management is not renewable resource management, the stated aim of most kinds of commercial forestry. It is not even farming.

It is mining.

VII

External Threats to Ancient Forests

Acidic Deposition and Tropospheric Ozone

Into my heart on air that kills
From yon far country blows. . . .
(Alfred Edward Housman, "To an athlete dying young," *A Shropshire Lad* [1896])

In a simpler world, the forester's task would be formidable: juggling the interests of hunters and timber producers, municipal water departments and salmon netters, and hikers and recreational vehicle users—each having ideas of how forestlands should be managed and seeing others as opponents to their interests. But as world population rises and human activities intensify, the forester's task is complicated by a new threat, one that past forestry classes did not address.

The composition of our atmosphere is changing. Not in a big way, mind you: The two gases that make up more than 99 percent of the atmosphere, nitrogen and oxygen, are hardly changing at all. But the remaining fraction is another matter. Certain trace gases (some naturally found in the atmosphere, others newly created by industrial pro-

220

cesses) are increasing. Their increase is already profoundly affecting some forests, and those effects are going to spread in coming years. Everyone—loggers and environmentalists alike—will lose if we cannot overcome this threat to our forests.

Several dozen gases are involved, each with a suite of origins and effects. The causes of their increase are diverse, including the cutting and burning of forests, the mining of deep coal deposits, the ranching of cattle in the tropics, the draining of marshes for agriculture, the growing use of air-conditioning, and, most of all, the increasing use of fossil fuel. The effects of these gases and their reactions in our atmosphere are exceedingly complex. It is beyond the scope of this book to provide an exhaustive look at atmospheric changes, but they are so relevant that no book on Westside forests can afford to ignore them.

The best-known atmospheric changes that affect forests are acidic deposition and tropospheric ozone, which have been studied in Europe and eastern North America for several decades. Although there are uncertainties about their relative importance in various places, it is clear that they are having severe effects on a large and growing fraction of the biosphere, particularly coniferous forests.

Acidic deposition—acid rain, acid snow, acid fog, and dry deposition of acidic particles—is caused by chemicals that enter the atmosphere when fossil fuels are burned and when metal ores are smelted. Coal, oil, and many metal ores contain sulfur. Sulfur from fossil fuels burned in electric power plants, cars, and home furnaces and from sulfide ores combines with oxygen to form sulfur dioxide, then with atmospheric water to become sulfuric acid.

At the high temperatures in automobile engines and other fossil fuel-burning systems, some atmospheric nitrogen combines with oxygen to form nitrogen oxides, which react with water to form nitric acid. Sulfuric and nitric acids in the atmosphere are then deposited across the land as dust, fog, or precipitation. Some fog and rain are more acidic than lemon juice, and many are acidic enough to dissolve marble, damage vegetation, and affect the chemistry of soils, streams, and lakes.

Acidic precipitation is only part of the problem. Nitrogen oxides and hydrocarbons that enter the atmosphere—mainly as emissions of unburned fuel from service stations, car, and truck engines—undergo complex chemical reactions in the presence of sunlight. The products of those reactions combine with oxygen (O_2) in the lower atmosphere (the troposphere) to form ozone (O_3), a highly reactive gas that attacks such diverse materials as rubber, lung tissue, and vegetation.

Some effects of acidic deposition and ozone on plants are similar, such as slowed growth, increased susceptibility to diseases and pests, and outright death. And since their causes overlap and many areas have both acid rain and ozone problems, it is difficult to distinguish their contributions to the *Waldsterben* (forest death) or forest decline affecting coniferous forests in much of Europe, the San Bernardino Mountains of California, the eastern United States, and Canada. Because prevailing airflow is from the west, even "pristine" eastern areas are experiencing serious damage to forest ecosystems.

Fortunately for the Northwest, very little sulfur oxides, nitrogen oxides, or hydrocarbons have been added since the air passed over industrialized areas of Asia. And since then, it has also mixed with cleaner air and been cleansed by precipitation during its passage. As a result, incoming air is unusually clean. Vancouver, Seattle, Tacoma, and Portland and various industries outside the cities add their combustion products to it, but Westside air quality is usually good. However, as the region's population grows, air quality will inevitably deteriorate unless strong measures address the increasing pollutant load. Few, if any, other north temperate forestlands are as safe from these pollutants.

Robert Edmonds and F. A. Basabe (1987) at the University of Washington have begun studying air pollutant levels and their effects on Westside forests downwind of population and manufacturing centers. They have found that summer ozone levels in some areas can reach levels known to harm sensitive trees; they have also observed damage in Douglas-firs.

By and large, the Westside does not get very acidic rain, although acid fog can occur in winter. Still, the effects of sulfur oxides are clearly visible by looking at biological indicators: lichens. As European scientists have known for decades, lichens are sensitive indicators of atmospheric sulfur oxides. Bill Denison and Sue Carpenter (1973) have used tree-dwelling lichens to map pollution levels in the Willamette Valley. Not surprisingly, the fewest species are found near large cities and towns, but the most sensitive lichens are absent even from some places far removed from major pollution sources.

Nonetheless, because prevailing airflow is so clean to begin with, acidic deposition and tropospheric ozone threaten Westside forest ecosystems much less than they do other forested regions. But by no means are we "out of the woods." The global atmosphere is changing in other ways that will have profound effects on Westside forests.

Stratospheric Ozone Depletion

In a meeting last week of the Cabinet council on domestic policy [Interior Secretary Donald] Hodel argued for an alternative program of "personal protection" against ultraviolet radiation, including wider use of hats, sunglasses and sun-screening lotions.
(Cass Peterson, "Administration Ozone Policy May Favor Sunglasses, Hats," *Washington Post*, May 29, 1987)

If an award were given for the most confusing atmospheric chemical, ozone would win easily. Ozone chemistry is bewilderingly complex, with two distinctly different effects on living systems, depending on where the ozone is. In the lowest layer of the atmosphere (the troposphere), human activities are *increasing* ozone concentrations in many places to a point where they are harmful to human health, crops, and natural ecosystems, including forests. But higher up, in the stratosphere, human activities are *decreasing* ozone concentrations. Ironically, stratospheric ozone depletion is also harmful.

Solar radiation includes both the visible light that fuels photosynthesis and wavelengths—especially ultraviolet-B (UV-B)—that are harmful to life. Ozone in the stratosphere, from ten to thirty miles up, is an efficient UV-B absorber. Depletion of this ozone eliminates the Earth's natural "sun screen," so the amount of UV-B reaching the Earth's surface increases. Although most media coverage has concerned profound stratospheric ozone depletion over Antarctica, to a lesser extent it is happening in temperate latitudes as well, including the Pacific Northwest.

There is normally an equilibrium between processes that create and break down stratospheric ozone. Now the balance has been tipped toward depletion by increasing atmospheric concentrations of nitrous oxide and (especially) chlorofluorocarbons (CFCs).

CFCs are synthetic chemicals that are emitted by aerosol spray cans (except in the United States, Canada, and Scandinavia, where use of CFCs as propellants has been banned), leaking refrigerators, and air-conditioners, and during manufacture of computer chips and some expanded foam packaging. CFCs break down in the stratosphere, starting a chemical chain reaction that breaks down large amounts of ozone. Ozone depletion is exposing the world's people and ecosystems to increasing amounts of UV-B radiation.

UV-B disrupts DNA. In humans, it damages the immune system and causes cataracts and skin cancers. It also harms other animals. For more than a decade, scientists have known that UV-B can harm herbaceous plants such as soybeans. Some conifers appear to be vulnerable as well, even at UV-B exposures only slightly above natural levels.

Susan Kossuth and R. Hilton Biggs (1981) found that elevated UV-B levels altered the growth of all but one of the conifers they tested. Biomass growth was reduced in southeastern loblolly and slash pines, Eastside ponderosa pines, and Westside lodgepole pines and noble firs. It was unchanged in Westside Douglas-firs and increased in Eastside white firs. Other measures, such as leaf area and root growth, were altered as well. Unfortunately, their results are difficult to interpret because the plants were grown at low visible light levels.

More recently, Joe Sullivan and Alan Teramura (1988) of the University of Maryland found that height and biomass growth are reduced in seedlings of some conifers from low elevations, particularly loblolly pine, but biomass growth actually increased in two high elevation species, an eastern fir and Utah Engelmann spruce, a species that also occurs in a few higher, drier parts of the Westside.

Unfortunately, we do not know how increased UV-B will affect Westside conifers. There has been little research, and the experimental methods have differed in key ways. Further, the published studies have been on seedlings in growth chambers or greenhouses; effects on older trees in intact ecosystems could be quite different. Scientists are years from predicting effects with reasonable confidence. However, we do know that the gases that deplete stratospheric ozone contribute to another effect, one better understood, more dangerous, and undoubtedly far more difficult to control: the greenhouse effect.

The Greenhouse Effect: Global Climatic Change

Global warming is a forestry issue.
(R. Neil Sampson, "Releaf for Global Warming," *American Forests* 94 [11–12] [1988])

The windswept, frozen Arctic tundra . . . the sunbaked, parched desert . . . the dripping, lush temperate rainforest . . . of all the factors shaping the Earth's patterns of biological diversity, stability, and productivity, climate is the most important. Climate determines the kinds of organisms that can live in an area, their abundance, their life spans, their rates of reproduction, growth, and death, and the outcome of their interactions with other species.

Climate also affects organisms on a regional and local scale. The striking differences between the Olympic rainforests and the mixed evergreen forests of northwest California reflect their differing climates. Differences nearly as dramatic can occur within just a few miles where topography causes sharp differences in temperature and rainfall patterns. Climate affects every aspect of forestry.

Of course, climate naturally varies in time. At times during the 3,500-million-year history of life, the Earth has variously been hotter or colder than it is now. The Northwest's climate and biotic communities have changed markedly in the last 2 million years as glaciers advanced and retreated. But in the last 6,000 years, certainly since Europeans arrived, the Westside's climate has been quite stable. Today's ecological patterns, human settlements, and economic activities are closely tied to recent climatic patterns.

This stability is about to end. Our planet's climate is beginning to change in ways that will affect sea level, biotic communities, fisheries, agriculture, forestry, cities and towns, transportation, and energy use— virtually every aspect of life on Earth. The greenhouse effect could bring warming in the next century as great as any since the last ice age. Further, these changes will occur with unprecedented speed. Although general circulation models do not yet agree *how* climates will change in all regions, no place on Earth, not even the Pacific Northwest, will be untouched. Most places will get warmer. Some will be wetter and some dryer, although climatologists are much less certain about future precipitation patterns than future temperature patterns.

Before we can gauge how climatic changes will affect Westside forests, we need the best possible picture of what those changes will be. I asked John Firor, Director of the Advanced Study Program at NCAR, one of the world's leading climate research institutes, to provide this foundation.

Global Climate Change: Some Basics

□ *John Firor, National Center for Atmospheric Research, Boulder, Colo.*

One of the better understood features of the Earth's atmosphere is the way in which the surface of the Earth is kept much warmer than our distance from the sun would ordinarily allow. Heat arrives at the Earth from the sun, mostly in the form of visible light. Some heat is reflected by clouds, ice caps, and other light-colored features; the rest is absorbed by the atmosphere and surface. To balance this input of energy, the Earth's surface radiates away infrared radiation. But the path of this energy back to space is complicated by the presence in the air of small amounts of gases that can absorb infrared radiation and send it back to the surface. This redirected radiation keeps the Earth's surface warmer than it would be otherwise.

The size of this heat-trapping effect is large. Were we able to remove all infrared absorbers from the air without changing anything else, the Earth's surface would be 33° C (59° F) colder; the oceans would freeze, and the Earth would be ice-covered.

So, the existence of an active and complex biosphere on Earth depends on the heat-trapping properties of the atmosphere. But this obviously beneficial effect also raises a concern. The amount of carbon dioxide, the leading heat-trapping gas in the atmosphere, is increasing rapidly as a result of fossil fuel use and deforestation. The concern, then, is this: If the historical amount of infrared-trapping gases in the air heats the Earth by a large amount, will adding more such gases to the air heat it even more? Unless there is some special and unknown effect that would stabilize the Earth's temperature—an effect that did not exist during hot geological periods in the past—the answer must be yes.

This general conclusion—that the Earth will warm as extra heat is trapped by added CO_2, methane, CFCs, nitrous oxide, and other infrared-trapping gases emitted by human activities—cannot yet be translated into precise predictions of temperatures and rainfalls in each region on Earth and at each date in the future. But the possibilities foreseen, when complex mathematical models are used to study future climate conditions, range from rates of change that would be troublesome and expensive to adapt to, to values that would clearly be catastrophic. The consensus expectation is that within a decade or two the global average temperature will be increasing at about 0.3° C per decade, some five to ten times faster than the changes seen during the two little ice age episodes between A.D. 1300 and A.D. 1800 or during the Medieval Warm Epoch in the eleventh and twelfth centuries. If the West Coast fully shares in this rate, the climate in Seattle will "move south" at 2 miles a year. The sea level rise accompanying the heating would accelerate to 5 cm (2 in.) per decade, up from the 1 cm per decade experienced in the last 100 years.

These numbers have caused international gatherings of experts to call for steps to slow the changes to allow human societies time to adapt in an orderly and planned way to altered living conditions, coastal activities, agriculture, and forests throughout the world. Slowing the change requires taking all available steps to slow deforestation, to increase the efficiency with which fossil energy is used, so the total use can be reduced, and accelerating the search for inexpensive alternative energy sources that can achieve public acceptance.

But some heating is inevitable; the gases already in the air can increase the temperature another 0.5–1.0° C beyond the warming of the last century, and the momentum in the use of fossil fuels will make decreasing their use a time-consuming job. The current state of the art foresees that the high latitude regions will heat more rapidly than the average and the equatorial regions less, that the interiors of mid-latitude continents will experience a loss in soil moisture while global rainfall will increase, and that winters will heat more than summers. The Westside could well be hotter than now, especially in winter, and perhaps wetter, but more refinements are needed in the ability to anticipate regional change to help us plan a response in each locality. The models cannot yet eliminate the possibility that this region will heat rather more or less than the average for that latitude or that some drying will occur.

The improvements needed in the models will not come easily. Larger computers are needed to perform the calculations on a finer scale than is currently possible, and a better understanding of the relationship of climate on smaller scales with topographic and other regional conditions is required to make full use of the added computer power. In addition (and especially for a region such as the Pacific Northwest), a better description of ocean currents, their effect on sea surface temperature, and their changes in response to a hotter climate will be necessary before much better guidance can be given to those planning regional adaptation to climate change. □

Climatic Change and Ancient Forests

[The] Earth's climate does not respond in a smooth and gradual way; rather it responds in sharp jumps. These jumps appear to involve large-scale reorganizations of the Earth system. If this reading of the natural record is correct, then we must consider the possibility that the major

responses of the system to our greenhouse provocation will come in jumps whose timing and magnitude are unpredictable. Coping with this type of change is clearly a far more serious matter than coping with a gradual warming.
Wallace S. Broecker ("Greenhouse Surprises," *The Challenge of Global Warming* [1989])

One autumn I gave a peach tree to friends in Vermont. It was a variety specially developed in New Hampshire for exceptional cold-hardiness, and should have endured most winters. But, by chance, that winter was unusually cold. When spring came, no leaves burst from the little sapling.

When you see a giant conifer 500 or 1,000 years old, you can be certain that it has experienced years much wetter, drier, colder, and warmer than average. It has probably survived for two reasons. One is exceptional tolerance of short-term fluctuations in weather. Trees that could not endure nature's capriciousness failed to pass on their genes long ago. The second reason if that within large stands, climate varies far less than in the surrounding areas. By standing together, trees create conditions that favor their perpetuation; that is, ancient forests display a kind of ecological stability called "resistance."

But climatic changes in the next century might well prove to be an irresistible force, so it is reasonable to ask whether forest species can be "moved" evolutionarily. We know that some species evolved in response to past climatic changes, but the extraordinarily high rate of future climatic change will make this impossible for many species. Because rates of evolution depend partly on generation time, bacteria (with generation times measurable in tens of minutes to days) and house flies (with generation times of several weeks) may respond fast enough. But evolution would proceed much too slowly in organisms with generation times of decades, such as trees.

Organisms can also respond to climatic changes by moving in another sense—by shifting their distributions—so you might ask whether the forests might not simply move northward. The answer is no; there will undoubtedly be movement, but it will hardly be simple.

The simplest response to changing climate would occur if the northward shift in isotherms—lines of equal temperature—were slower than the dispersal rate of the slowest dispersing species. For this to happen, the ecological communities of today would have to move northward at about two miles per year. The species of southwest Oregon, for example, would have to arrive in southwest Washington in about 150 years. Looking at it the opposite way, we estimate that every spot in the Northwest would, in effect, "move south" two miles a year.

This simple picture of shifting ecological zones is likely to be wrong for many reasons. First, a shift of two miles a year is probably within the capabilities of species with high dispersal rates and short generation times. These are often the species we call weeds and pests. Beside migrating quickly enough to keep up with the climatic shift, they also might evolve quickly enough to deal with environmental changes further north. One such change that few people (other than chronobiologists) think about is daylength.

Daylength changes with latitude: Between the beginning of spring and the beginning of summer, for example, daylength increases faster with increasing latitude. Because daylength correlates with the time of year more reliably than current weather, behaviors of many animals and the phenology (the timing of yearly events, such as when to flower) of many plants are tied to daylength.

Consider a plant species living at Grants Pass, Oregon, that starts photosynthesizing when air temperatures reach 5° C (41° F), which can happen any time in the early spring, but flowers only when the daylength reaches fourteen hours, which occurs predictably on April 30. If climatic change forced this plant's progeny to migrate to Olympia, Washington, they would experience a fourteen-hour day on April 22. Assuming the same climate for the Grants Pass of today and the Olympia of tomorrow, they would have an average of eight fewer days to store photosynthate before flowering. Less energy would be available for producing nectar, pollen, and seeds, which would diminish their reproduction. This could hamper their ability to compete with other plants and recover from setbacks such as defoliation and bad weather.

Species with short generation times might evolve fast enough to adapt to the new photoperiods, perhaps by resetting their internal clocks. Such species are least likely to be harmed by climatic shifts. Pests and weeds should not suffer too much.

Species with long generation times—such as trees—would have more difficulty. Displaced to a site with the same climate but a new photoperiod, their genes might lock them into inappropriate phenologies, such as producing new needles early in the season when they are vulnerable to late frosts. This would disadvantage them in competition with weedier species.

But many long-lived species would face a larger problem. The rates at which they can disperse, although faster than their rates of evolution, are still far slower than anticipated climatic changes.

During the last glaciation, trees that now dominate boreal forests in upper Michigan and central Ontario occupied the southern United States. After the glaciers began to retreat, various tree species moved northward from their refuges at different rates, with some still migrating after 15,000 years, a rate far slower than the rate at which climate changed. Tree species advanced an average of perhaps 300 meters per year, about ten times slower than they would need to move to keep pace with advancing isotherms (lines of equal temperature) in coming decades.

If Westside conifers move at roughly comparable rates, individuals at the southern part of their distributions will be overtaken by warmer zones. As they are pushed past their physiological tolerances, they will cease reproducing successfully, then cease growing, and finally, when conditions become stressful enough, succumb. And although trees are probably better adapted to withstand normal weather fluctuations than most living things, it seems unlikely that they will be unscathed by temperatures that will exceed any in the last 100,000 years.

For species with naturally limited ranges, such as Port Orford cedars and Brewer spruces, all individuals could well be overtaken by unsuitable climates, which would cause extinction. But shifting climatic zones would be severely damaging even to trees with larger ranges that are highly adapted to conditions where they live: species with ecotypic variation, such as Douglas-firs and sugar pines. Large mortalities could occur, eliminating important genetic diversity. At the very least, their ranges would shrink in the south far more than they would expand in the north.

Another problem that species will face on their forced march is a change in soils. Climate affects soil formation by determining the rates of weathering of parent material, leaching (downward movement of ions in soil water), and decomposition of organic material. The distribution of soils is a function of climate.

But the rate at which soils evolve would probably lag behind the rate of climatic change. After hundreds or thousands of years, soils would probably reach equilibrium with the new climate. But until then, they would pose a novel situation for newly arriving plant species. By luck, some plants might do better, but many more would likely fare worse in soils to which they are not adapted.

Soils vary geographically for another reason: They are derived from parent materials that differ from one place to the next. Some species are adapted to soils with unusual chemical compositions, such as those derived from serpentine rock. Serpentine is inimical to the growth of most plants. Its pH is often too acid or too basic, and it usually has low concentrations of nutrient elements including calcium, nitrogen, phosphorus, potassium, and molybdenum, as well as has high concentrations of toxic elements such as nickel and chromium. For these reasons, serpentine soils support unusual communities of plants, many of them endemics. These plants either require the peculiar chemical mix in serpentine soils or are tolerant of them but poor competitors with other plants on less stressful soils.

Serpentine species, such as Jeffrey pine, Port Orford cedar, Baker's cypress, Siskiyou butterflyweed, and Howell's fawn lily, would face special problems as climatic zones shifted. Outcroppings of serpentine soils are discontinuous, like islands. They are common in the Klamath and Siskiyou mountains, but the nearest Westside serpentine soils to the north lie some 400 miles away in Snohomish County, Washington. If serpentine species cannot bridge the huge gulf between southwest Oregon and northwest Washington, they will be unable to escape as the climate changes. Extinction of many species is likely.

But the problem of barriers to dispersal goes far beyond plants adapted to odd soils. Many other species would encounter natural barriers to northward movement. The best-known barrier is the Columbia River. Because it is so difficult to cross, species such as dusky-footed woodrats, red tree voles, western red-backed voles, and gray foxes have their northern limits at the Columbia.

Again, by looking to the past, we can gain some idea of what the future has in store. Before the Pleistocene ice ages, Europe was about as rich in tree species as North America; now it is much poorer. Paleobotanists believe that the Alps, Pyrenees, and Mediterranean Sea, which run east-west, prevented many species from migrating southward and northward during climatic shifts. Unable to escape, many trees became extinct. In North America, most natural barriers run north-south, so climatic shifts eliminated fewer species. The Columbia is an exception, but there are new barriers that did not exist in the Pleistocene, as we will see later.

Species stressed by warming have an option besides evolving or moving northward. As Rob Peters and Joan Darling (1985) discuss, they can also move upward because higher elevations are cooler. We know that some formerly widespread species found refuge in mountains when climates warmed in the past. Will this happen again? The answer is probably yes, to some degree, but there is an unprecedented problem.

High elevations are not the same as high latitudes in one crucial way: UV-B radiation levels *decrease* gradually as latitude increases but *increase* sharply as altitude increases. True, species seeking refuge at cooler elevations faced higher UV-B levels in the past. But the amounts of UV-B they encounter will be higher in the future because of stratospheric ozone depletion, as mentioned earlier. Low-elevation species could encounter damaging UV-B levels in refuges from greenhouse warming. This would not improve prospects for their survival.

Greenhouse warming is likely to bring both warmer winters and warmer summers, with different effects. Warmer winters would prolong the growing season, which, on first view, might seem to be good news. But as Dick Waring and Jerry Franklin (1979) suggest, conifers dominate broadleaf trees in much of the region because, unlike broadleaves, they can maintain their photosynthetic tissues at below-freezing temperatures. Warmer winters would remove this advantage, expanding northward the area in which tanoak, madrone, and golden chinquapin can outcompete conifers. This would not likely gladden the hearts of those who consider these (or any plants but Douglas-firs) to be "weeds."

Warmer winters would decrease the proportion of precipitation falling as snow and melt winter snowpacks earlier, thereby lengthening the period of low summertime streamflow. This would affect everything that depends on streams and rivers: salmon fisheries, the wildlife that concentrates in riparian corridors, ecological communities on floodplains, water-based recreation and agricultural, industrial and municipal water supplies.

Summer temperatures are no less important. Increased temperatures would increase transpiration; trees would need more water. In effect, ecosystems would become drier. Riparian and floodplain communities already affected by diminished summer streamflow would be particularly vulnerable.

Further, today's distributions of populations, species, and ecosystems reflect both average weather and deviations from those averages. Ecologists have observed that many long-lived organisms, especially those near the edges of their ranges, do not reproduce successfully in many years. Either they do not produce young or seedlings or their progeny do not survive. Rather, individuals are recruited into the population only during years or strings of years having conditions suitable to reproduction and survival of early life history stages.

Intriguing evidence of this comes from the widespread difficulties in reforesting clearcuts. Replacing the original forests (or any forests at all) is difficult on dry aspects throughout the Westside and is especially so in southwest Oregon and northwest California. The likely reason is that dominant trees can survive as established adults, but their seedlings cannot compete successfully with other, better-adapted colonizers of hot, dry habitats (such as burns and clearcuts) in normal years. They can establish, even with help from humans, only in especially favorable years. Under current climatic conditions, Douglas-firs, western redcedars, and western hemlocks might not establish successfully more than a few times per century in these areas.

For these reasons, ancient forests are unlikely to replace themselves in the face of climatic changes. While they resist change, they are not very "resilient" or able to recover after a disturbance if they are stressed. Because conditions for survival are broader than those that support establishment, a warmer world will pose serious problems. Especially on warm, dry sites, even slight warming could eliminate the rare, favorable years that forests need to reestablish.

However, anticipated warming is not likely to be slight, and warming will not be the only climatic consequence of increasing greenhouse gases. All of the general circulation models used to examine possible future climates suggest that warming will be greater toward the poles. The resulting changes in latitudinal thermal gradients will likely affect the seasonal locations of the high- and low-pressure areas whose interactions determine the Northwest's weather patterns. Any such shift would alter precipitation patterns, possibly including the total amount of precipitation, the length of the dry season, and the frequency, duration, intensity, and variability of rain, snow, and fog. It is premature to predict *how* these will change, but considering the sensitivity of Westside plant communities to moisture, it seems far-fetched that altered precipitation patterns would not affect forests in the next century.

One example illustrates how a possible consequence of greenhouse warming could bring major changes to the Westside. Coastal Sitka spruce/western hemlock communities (and coast redwood communities to the south) depend on fog. Fog has two effects on their water relations. By lessening the intensity of sunlight, fog reduces transpira-

Steven C. Whitney

Trees burned by the Silver Complex Fire, Siskiyou National Forest (Oregon). *This 1987 fire, like virtually all major natural forest fires, began after lightning ignited trees during an unusually hot, dry summer. Greenhouse warming will effectively make ecosystems drier by increasing water loss from plants. How it will affect rainfall and lightning frequency in the Northwest cannot yet be predicted accurately. Increased fire frequency caused by greenhouse warming could threaten both ancient forests and tree plantations.*

tion during rainless periods. And through fog drip, it provides added precipitation. Thus, it both lowers water loss and increases water gain.

Coastal fogs occur when warm, moist air from the sea invades cooler coastal land. *If* the greenhouse effect warms the land surface more than the sea, as current general circulation models predict, fog will be less frequent. This could eliminate Sitka spruce and coast redwood forests.

Greenhouse warming will likely cause other changes. Shifting high- and low-pressure cells would shift the direction and strength of winds, an important factor in a region where windthrow kills many trees. How winds might change is not yet safely predictable.

Further, thunderstorms—with attendant lightning—could become less or more common; again, there is no way to predict this with confidence as yet. Although Westside thunderstorms are rare, they are crucial because lightning causes virtually all natural fires, including the fires that swept more than 800,000 acres in the exceptionally dry summer of 1987, mainly in Klamath and Siskiyou national forests.

Fire regimes are highly sensitive to weather. Fires burn hotter and faster in the afternoon than in the morning. Westside fires occur mainly in the late spring through the fall, when hot, dry, windy weather can fan lightning strikes. And weeks, months, or years of warmer and drier than average conditions can set the stage for holocausts. It is not only average weather conditions but their variability that determines fire regimes. Even if average weather does not change, increasing interannual variability would increase the frequency of dry years, and thus increase fire frequency.

Thus, some climatic changes that *will* occur—higher temperature, higher evaporation and transpiration rates, and lower summertime streamflow—and some that *could* occur—lower humidity and precipitation, higher frequency of easterly winds, lightning, and annual variability—would increase the vulnerability of Westside forests to wildfires. Les Cwynar (1987) of the University of Toronto concluded from studies of plant communities of the northern Washington Cascades since the last glaciation that the most important effect of climatic change on Westside forests could be changes in the fire regime. As James Clark of the University of Minnesota recently observed, "There has been no serious consideration of recent climatic changes as a factor in fire regimes. . . . With continued fire suppression and the further warming, fuel build-up will result in more intense and/or more frequent fire" (1988, p. 234).

Climate also has profound effects on the survival and reproduction of insect pests and pathogens. Without knowing precisely how climate will change, it is premature to predict precisely how they will be affected. We do know that defoliating insect outbreaks are far less frequent and severe in Westside forests than in the harsher Eastside forests, and preliminary findings by Dave Perry and G. B. Pitman (1983) suggest that Westside conifers have little or no chemical defense against western spruce budworms, which are important defoliators on the Eastside. If the climate that, until now, has deterred pest outbreaks changes to one that allows defoliator outbreaks in Westside forests, these pests will encounter essentially no resistance. As important as direct climatic effects might be, changes in fire and pest regimes might be even more so.

Despite all this, J. Laurence Kulp (1986) and James Woodman (1987) both feel that anticipated climatic changes are unlikely to have an effect or would actually "help" timber production. Even if timber production could be equated with the health of forest ecosystems, their optimism would seem premature. Under the region's fastest rotations—

forty to fifty years—the highly selected trees being planted today will not be adapted to the climate that will prevail when they are scheduled for harvest. On the less productive national forest and BLM lands, where trees grow more slowly and rotations are longer, changing climate is still more likely to affect yields in the current rotation.

Military historians have observed that nations wedded to the past usually prepare to fight their previous war, much to their detriment, as when France depended for its national security on an impregnable Maginot Line of fortresses built in the path of an earlier German invasion. Unfortunately, in 1940, Hitler's armies simply went around the fortresses, and a defenseless France was forced to surrender in a few weeks. Considering the frequent failure to reforest cutover sites even under current, known conditions, planting trees designed for a world that will no longer exist suggests that tree plantations could become the Pacific Northwest's Maginot Line.

Direct Carbon Dioxide Effects

Fair is foul, and foul is fair:
Hover through the fog and filthy air.
(William Shakespeare, *Macbeth* [1605–06])

Unlike other trace gases that deplete the ozone layer and intensify the greenhouse effect, CO_2 has direct physiological effects on plants. On first glance, these effects should be welcome.

CO_2 is one of the two chemical raw materials (the other is water) for plant photosynthesis. Insufficient CO_2 sometimes limits plant growth; increasing the concentration of CO_2 increases growth rates in many temperate zone plants, an effect called "CO_2 fertilization." CO_2 also benefits plants' water relations. Plants take CO_2 in through microscopic pores (stomata) found mostly on leaves. But open stomata also allow water loss (transpiration). At higher CO_2 concentrations, plants can keep more stomata closed, so they lose less water.

If, perchance, CO_2-induced greenhouse warming does not change total precipitation one way or the other, it will still affect Westside ecosystems in ways that effectively make them drier during the growing season, as discussed in the previous section. The silver lining to this cloud is that elevated CO_2 will also lessen plants' water losses, to some extent softening the impact of warmer conditions.

Responses to increased CO_2 have been tested for a number of plants (mostly herbaceous ones). Seedlings of southeastern loblolly pine and sweetgum (a broadleaf tree) show sharply increased growth rates. Sweet-gums also show reduced water loss, but loblolly pines and Sitka spruce do not.

Whether the goal is faster wood production or faster replacement of degraded wildlife habitat, increased plant growth rates and improved water relations might cheer timber producers and conservationists alike. But before opening the champagne to toast this rare good fortune (thereby releasing more CO_2 into the atmosphere), it is wise to ask, "What's the hitch?" Unfortunately, there could be at least two.

The pioneering research by Boyd Strain (1987) and colleagues at Duke University shows that plant species respond differently to increased CO_2 levels, which enhance growth of some more than others. Differential growth (and, presumably, reproduction) would alter competitive relationships among plant species, causing changes in community structure, species composition, and ecosystem processes such as nitrogen fixation and soil formation.

If species that people like (such as trees favored by the timber industry) become more competitive relative to other plants, many people will welcome these changes. However, they could also become *less* competitive. Species that vegetation managers consider "undesirable" now could become truly serious weeds in the next century.

Further, CO_2 enrichment decreases the nitrogen content of plant tissues, to which plant-eating insects respond by increasing their feeding rates. From the first such study on possible effects of herbivorous insects on plants grown in CO_2-enriched atmospheres, D. E. Lincoln and coauthors concluded that "the increased levels of plant productivity at higher CO_2 concentrations may be offset by higher herbivory and could even be reduced below current levels" (1984, p. 1529). More recent research on four other herbivorous insects and their hosts has consistently found increased feeding rates.

It is premature to extrapolate these results to Westside trees in their ecosystems. Further, since experimentally testing the growth of large trees in enriched CO_2 atmospheres will be difficult, expensive, and time-consuming at best, we are not likely to have a good idea of which forest species will gain at the expense of others anytime soon. But the direct effects of CO_2 might be important enough that they need to be included in any assessments of future forest development. How they will affect forest ecosystems also being exposed to increased UV-B radiation and direct and greenhouse warming is anyone's guess.

When Trends Collide: Timber Operations and External Threats

Habitat destruction in conjunction with climate change sets the stage for an even larger wave of extinction than previously imagined.... Small remnant populations of most species, surrounded by cities, roads, reservoirs, and farm land, would have little chance of reaching new habitat if climate change makes the old unsuitable. Few animals or plants would be able to cross Los Angeles on the way to the promised land.
(Robert L. Peters, "Effects of Global Warming on Biological Diversity," *The Challenge of Global Warming* [1989])

Timber operations are changing the species composition, structure and functioning of Westside forests. At the same time, human activities will bring rapid climatic change. These effects could either counteract or amplify one another. What will happen as trends collide?

There are uncertainties about cumulative effects of timber operations, larger uncertainties about regional climatic change, and still larger uncertainties about effects of increased UV-B radiation and atmospheric CO_2 levels. Therefore, most specific predictions about their interactions are speculative. Still, some general themes are becoming clear.

Timber operations are producing landscape patterns that maximize edge between ancient forests and surrounding patches. Under current rotations, few (if any) surrounding patches will ever reach the height of ancient forests. The sharp contrast between remnant ancient forest patches and surroundings increases risk of blowdown. Greenhouse-induced changes in windstorms could either lessen or amplify this risk.

As fragmentation increases, forest edge increases rapidly and forest interior decreases rapidly. As edge increases, the harsher surrounding conditions penetrate an increasing fraction of ancient forest and the area

Clearcut, Olympic National Forest (Washington). *The climate in a clearcut is more severe than that of the ancient forest it replaces: hotter, colder, drier, windier, and with greater snow depth. Summer winds from clearcuts penetrate nearby forests, making them hotter and drier. This harms species that need cool, moist conditions, and makes the forest fragments more vulnerable to fire.*

of forest interior diminishes. Thus, timber operations are changing the climate within ancient forests. Greenhouse warming (which is also accelerated by logging ancient forests) will amplify these changes, a potentially disastrous synergism.

Some climatic effects will be direct. For example, the species that need cool, moist conditions (including salamanders, slugs, and mushrooms) are particularly vulnerable. So are aquatic species that need cool water or high concentrations of dissolved oxygen. Although buffer strips of trees now provide some shading for larger streams, small streams (which are major habitats for aquatic amphibians) do not get the same protection. Deprived of cover by timber operations, they heat in the sun. The amplifying effect of greenhouse warming will not be good for species that need the cool, well-aerated conditions of ancient forest headwaters.

There will also be indirect effects. A major one is an altered fire regime. As the last remnants of forestland are roaded, the number of ignition sources from escaped slash burns, chainsaws, campfires, and carelessly tossed cigarettes will increase. As clearcuts and young tree plantations replace ancient forests, a growing fraction of the landscape will have fuel distributions that favor the rapid spread of fires, especially crown fires. Together, these factors are likely to increase the impact of wildfire in Westside landscapes. Amplifying them will be the effects of greenhouse warming.

Greenhouse warming will cause earlier snowmelt (thus increasing the length of the dry season) and higher transpiration, effectively making ecosystems drier. How precipitation will change is a key unknown. A major increase would help counteract increased fire risk from timber operations and warming. A modest increase could fail to offset them. If there is no change, or if precipitation decreases (as some models suggest), fire frequency will rise sharply; the severe fire seasons of 1987 and 1988 could just hint of things to come. And although fire frequency and severity are now inversely related, climatic changes could increase *both* fire frequency and severity until species and fuels reach equilibrium with the new fire regime.

Timber operations and changing climate might have similarly reinforcing effects on insect and disease outbreaks. Unlike Eastside forests, Westside forests have been little troubled by native defoliating insects such as western spruce budworms. The insects are present in low numbers, but, as James Hadfield (1988) of the Forest Service points out, the cool, moist climate prevents population outbreaks by making their larvae vulnerable to predators and parasites.

But the landscape once dominated by ancient forests (whose heterogeneous structure, diverse predators, and microclimate resist the spread of pests and diseases) is becoming dominated by tree plantations (which promote their unchecked build-up and spread). At the same time, the climate is getting warmer.

As farmers and epidemiologists know, pressures from insect pests and pathogens are generally higher in warmer climates, and stressed organisms (such as heat- or drought-stressed conifers) are usually more vulnerable. It would be very surprising if warming did not exacerbate pest outbreaks in Westside forests.

There are striking parallels between projected changes in fires and in insect pests and diseases. In many cases, greenhouse warming and alterations of landscape patterns by timber operations will have additive or synergistic effects on these agents of disturbance. Westside forests will probably be affected by climate-induced changes in the disturbance regime long before warming harms trees directly.

Some changes in temperature-dependent ecological processes will appear to be beneficial. A principle of physiological ecology is that an increase of 10° C increases the rate of biological processes by about 100 percent. If, for example, greenhouse warming and timber operations together increase temperatures on a site by 7° C, the decomposition rate would increase by about 50 percent. Because woody debris and duff tie up nutrients that can limit plant growth, accelerated decomposition will liberate nutrients faster.

However, higher temperatures will accelerate decomposition of snags and logs, diminishing their longevity and, thus, their abundance, further jeopardizing species already hard-pressed because timber operations are eliminating these essential habitat components.

Furthermore, more rapid decomposition will affect greenhouse warming. Ancient forests store unsurpassed amounts of carbon because their rate of primary production—the amount of CO_2 they fix—is high while the death rates of big trees and decomposition rates are low. How altered climates will affect primary production is unclear. But the faster the organic matter decomposes, the more CO_2 will be liberated. Warming caused by both timber operations and the greenhouse effect will further hasten carbon release, which will increase warming still more. This is a positive feedback—a "vicious cycle"—that we would do well to avoid.

Climatic changes will affect species differently, with short-lived, fast dispersers (such as weeds and pests) least likely, and long-lived, slow dispersers (such as giant conifers) most likely, to be harmed. Warming will force species northward, but some species will be unable to move fast enough and will vanish. Others will be stopped by barriers such as the Columbia River and those created by humans: cities, farmlands, roads, tree plantations, and clearcuts.

Timber operations have created barriers to dispersal by fragmenting natural forests. Many ancient forest species—from jelly fungi and slugs to Pacific yews and spotted owls—are unlikely to disperse well across cutover landscapes. Their problems will increase with increasing contrast between ancient forest and surrounding areas and with increasing distance between suitable ancient forest patches.

Pushed by changing climate yet stopped by barriers caused by logging, many species will encounter severe problems shifting northward. But there is a ray of hope. Mountains are naturally cooler than lowlands, and the Pacific Northwest is well endowed with them. Perhaps they will serve as refuges for species from the lowlands.

However, there are hurdles in the path of this upward shift. Smaller areas support fewer species and smaller populations. Because area decreases with increasing elevation, mountains can accommodate fewer species than lowlands.

Another factor could make mountains still less suitable as refuges. UV-B naturally increases sharply with altitude, but the world is entering an era of increased UV-B radiation because of stratospheric ozone depletion. As climate deteriorates in the lowlands and suitable temperature zones ascend, high UV-B levels could limit colonization of mountains by lowland species.

At the same time that these changes are occurring, CO_2 will probably be altering competitive relationships among plants by accelerating their growth in differing degrees and increasing the feeding rates of insects that eat various species.

It is difficult enough to prepare for the future when we are confronted with one change of great magnitude. But humankind is facing many major changes simultaneously, changes whose effects can reinforce one another. All are products of uncontrolled human population growth and resource consumption. None can be prevented; they are already under way, and the best we can do—and must do—is to limit the damage.

This will require much better understanding of our ecosystems—by scientists, educators, the media, businesses, decision-makers, and the general public—than we currently possess. Any action to lessen these changes will require unprecedented cooperation among the competing forces that created them. Can society provide scientists with the resources we need to map the safest paths through the minefield that awaits us? Can CFC manufacturers find substitutes that will not destroy the ozone layer? Can coal-rich nations summon the will to conserve energy and switch to energy sources that do not worsen greenhouse warming? Can the timber industry practice sustainable forestry instead of mining the ancient forests of their carbon and their wildlife?

A great deal is riding on the answers to these questions.

VIII

Sustainable Forestry for the Pacific Northwest

How Much Old-Growth Remains?

Management options that include retention of existing old growth must be given priority.
(Jack Ward Thomas et al., "Management and Conservation of Old-Growth Forests in the United States" [1988])

Before settlers arrived, most of the Westside was cloaked in ancient coniferous forest. The rest was mainly higher and colder subalpine forest, alpine tundra and glacier, or warmer and dryer grassland or oak woodland. Since then, forestland has been replaced by clearcuts, agricultural lands, roads, and municipalities. How much remains, and how fast is it disappearing?

The answers to these questions are central to any sound conservation or timber production strategy. Yet data are incomplete and difficult to come by. No comprehensive regional inventory or map of old-growth exists; existing inventories and maps for portions of the region are of limited value. Data from northwestern California are especially scarce, so this analysis concerns only western Washington and Oregon.

243

How much ancient forest was there? The land base of western Washington and Oregon is just less than 35 million acres. The proportion that was forested before Euro-American settlement must have been higher than today. Jerry Franklin and C. T. Dyrness (1973) cite an estimate that 82 percent is now forest; E. Reade Brown and Alan Curtis (1985) give 78.5 percent. The lower figure equals 27 million acres. According to sources in Larry Harris's study (1984), 85 percent of western Washington forests and 90 percent of western Oregon forests were considered old-growth, or 24 million acres. These percentages seem high; some probably would not have met the Forest Service's ecologically meaningful interim definition. Assuming that 20 percent would have failed gives an estimate of 19 million acres of ancient forest before settlement. Harris gave the same figure; Tom Spies and Jerry Franklin (1988) suggest 15 million acres. We will never know precisely how much there was, but the 19-million-acre estimate is at least a plausible starting point for assessing what has been lost.

How much is left? Two recent studies by Forest Service researchers Richard Haynes (1986) and Sarah Greene (1988) provide estimates (see Table 8.1). Despite large differences, they at least agree that most remaining old-growth is in the nine national forests: Mt. Baker-Snoqualmie, Olympic, and Gifford Pinchot in Washington; and Mt. Hood, Siuslaw, Willamette, Umpqua, Rogue River, and Siskiyou in Oregon. The Forest Service has issued draft management plans for all nine. These estimates are used for scheduling timber sales for the next fifty years. Thus, their accuracy is crucial to the future of the ancient forests.

Table 8.1

PREVIOUS ESTIMATES OF OLD-GROWTH FOREST IN WESTERN
WASHINGTON AND OREGON
(ACRES IN THOUSANDS)

Ownership Class	Haynes (1986)	Greene (1988)
National forests	1,876	3,347
National parks	660	766
BLM lands (in Oreg.)	305	478
Other public lands	47	—
Private lands	390	—
Total	3,278	4,591
Percentage of original old-growth	17	24

Differing definitions are one reason that the estimates vary. Haynes defined old-growth as stands over 250 years. The nine national forests from which Greene got her data used seven different definitions of old-growth in their draft environmental impact statements (EIS). Neither of these researchers stated how they derived these estimates. And there is another source of variation: The estimates are out of date. For example, Haynes based his estimates on timber inventories that are now more than a decade old.

Without reliable estimates, public debate and decision making on the future of ancient forests rest on shaky ground. To provide a rigorous, ecologically meaningful inventory, The Wilderness Society commissioned forest ecologist Peter Morrison (1988) to quantify how much old-growth remains in the national forests. Although The Wilderness Society had hoped to inventory twelve forests (including the three in northwestern California), resource constraints forced it to choose just six, all in Washington and Oregon. The ones it chose, however, span the region from cooler to warmer climes, from coast to Cascades, and allow cautious estimation about the status and trends of ancient forests throughout the region.

Morrison's study is the first to use the Forest Service's interim old-growth definitions. It applies them to data from Forest Service timber inventory plots in Mt. Baker-Snoqualmie, Olympic, Gifford Pinchot, Mt. Hood, Willamette, and Siskiyou national forests and updates them to 1988.

Because timber inventory plots had not been established in wilderness areas that existed at the time of the inventories, Morrison compared forest texture on aerial photographs of points in wilderness areas with those where timber inventory plots are in old-growth. A summary of Morrison's results appears in Table 8.2.

Morrison's estimates are lower than those from the national forest EISs and Haynes (Table 8.3). His data show that the six national forests do not have 2,543,000 acres of old-growth, as claimed in the EISs, but only 1,140,000 acres, only 45 percent as much. Haynes's estimates are consistently higher than Morrison's but lower than those in eight of the nine EISs (Tables 8.3 and 8.4), averaging only 63 percent as much.

Table 8.4 makes the assumption that the three remaining national forests in western Oregon inflated estimates as much as Morrison's six, deflating the estimates by using the above 45 percent figure.

However, these estimates of old-growth in the other three western Oregon national forests are probably still too high. They are the only

Table 8.2

OLD-GROWTH IN SIX WESTSIDE NATIONAL FORESTS IN 1988
(YEAR LAST INVENTORIED IN PARENTHESES; ACRES IN THOUSANDS)

Status	Mt. Baker-Snoqualmie (1976)	Olympic (1974)	Gifford Pinchot (1981)	Mt. Hood (1986)	Willamette (1981)	Siskiyou (1978)	Total
When inventoried	326	179	159	184	320	185	1,353
Mid-1988	297	106	119	178	299	141	1,140
<80 a. or within 400 ft. of clearcut by 1988	62	24	81	77	133	62	439
Loss/year	2.417	5.214	5.714	3.000	3.000	4.400	

SOURCE: Morrison (1988).

Table 8.3

ESTIMATES OF OLD-GROWTH IN MORRISON'S SIX WESTSIDE
NATIONAL FORESTS
(ACRES IN THOUSANDS)

National Forest	EISs	Haynes (1986)	Morrison (1988)
Mt. Baker-Snoqualmie	667	433	297
Olympic	217	205	106
Gifford Pinchot	231	232	119
Mt. Hood	346	218	178
Willamette	639	338	299
Siskiyou	443	171	141
Total	2,543	1,597	1,140

Table 8.4

ESTIMATES OF OLD-GROWTH IN WESTERN OREGON AND
WASHINGTON NATIONAL FORESTS
(ACRES IN THOUSANDS)

National Forest	EISs	Haynes (1986)	This Study
Siuslaw	34	10	15
Umpqua	619	219	278
Rogue River	150	50	67
Subtotal	803	279	360
Morrison's 6 forests	2,543	1,597	1,140
Total Wash. and Oreg. forests	3,346	1,876	1,500

ones that are higher than those in Haynes, who did the most rigorous
estimates before Morrison. Morrison's estimates for the other six forests averaged only 70 percent of Haynes's. Nonetheless, I will use the
higher of the two sets of figures.

Haynes's figures are too low, however, because he did not include old-
growth acreage in North Cascades National Park. Greene's figure of
766,000 acres is probably reasonable for the four Westside national
parks, so 106,000 acres should be added to Haynes's figure. Assuming
that Haynes's figures for other land ownerships are correct, western
Washington and Oregon do not have 3,278,000 acres of ancient forest,
but 3,008,000 acres. But if Haynes's bias is systematic across all other
ownerships, there are 2,796,000 acres: $1,500 + 766 + (1,140/1,597 =
0.714) \times (305 + 47 + 390)$. Assuming that all old-growth on private

lands is now gone (as most authorities suggest), subtracting a further 279,000 acres (390,000 × 0.714) gives 2,517,000 acres. These adjusted figures represent 16 percent, 15 percent, and 13 percent of 19 million acres. My best estimate is that 13 percent of the original ancient forest remains in western Oregon and Washington.

Some perspective is useful here. In recent years, destruction of tropical rainforests and U.S. wetlands has aroused the concern of scientists, political leaders, and the public, with good reason. Perhaps up to 15 percent of the ancient forest in the Amazon Basin of Brazil is already gone. About 50 percent of our country's original wetlands has disappeared. But as serious as these losses are, they are far less than the 87 percent of Westside ancient forest that is already gone.

The 13 percent still existing is unlikely to remain for long. In combination, the six forests that Morrison studied are losing 23,745 of their 1,140,000 acres per year (2.1 percent). Applying this figure to the three others in Washington and Oregon gives a total of 31,233 acres lost per year in the national forests alone. Evenly spread among these nine, this would mean the elimination of all the national forest old-growth outside existing wilderness areas in thirty-four years.

The situation is actually far worse. Small old-growth stands have a high edge:center ratio, which subjects them to windthrow and excessive external influence on their internal microclimates. Further, stands within 400 feet (roughly two tree lengths) of a clearcut or a road are often degraded when snags are cut for safety reasons. Combining these two categories in Morrison's six forests, 439,000 of the 1,140,000 acres are thus compromised, or 39 percent. Applying the 39 percent figure to the total of 1,500,000 acres yields an estimated total of 922,000 acres of uncompromised old-growth in the nine western Washington and Oregon national forests.

Even if none of the old-growth outside national forests is similarly compromised (an unlikely possibility where logging is allowed), there are only 1,939,000 (2,517,000–578,000) uncompromised old-growth acres. Assuming that none of the ancient forest in national parks is compromised, if 39 percent of the 251,000 (352,000 × 0.714) old-growth acres on lands managed by the BLM, the Washington Department of Natural Resources, and the Oregon Department of Forestry are compromised, the amount of uncompromised old-growth is 1,841,000 acres, or 10 percent of the original total.

Knowing how much is being lost in the national forests (Table 8.2) allows us to project dates (Table 8.5) when all the old-growth will be gone outside wilderness areas in each national forest ("Date All Un-protected O-G Gone") if recent trends continue. But these numbers are too optimistic for two reasons. First, because of edge effects, we really lose about 60 acres of old-growth for every 25-acre-patch that is clear-cut. Assuming that the Forest Service will continue to push into the last remaining roadless areas as quickly as possible and maximally disperse its clearcuts, multiplying the loss rate by 2.4 estimates how long un-compromised old-growth will last ("Date All Unprotected, Unfrag-mented O-G Gone"), a plausible worst-case scenario. Second, the years under "Date All Unprotected O-G Gone" assume that national forests are managed completely independently of one another. But as the last ancient stands are cut in some forests, pressure to compensate for their loss will likely increase logging in the others.

Table 8.5 shows that the last bastion of uncompromised old-growth in these six national forests will probably be Mt. Baker-Snoqualmie National Forest. Data in Greene (1988) suggest that old-growth is disappearing at 16,450 acres or 3.4 percent per year on BLM lands in western Oregon. Assuming that elimination of old-growth on state lands and nonwilderness lands in the Siuslaw, Rogue River, and Umpqua National Forests at least equals the 3.0 percent per year loss in Morrison's six national forests, uncompromised old-growth will cease to exist outside wilderness areas, research natural areas, and national parks in 2009, and the year 2062 would see the end of all unprotected ancient forest. But if cutting increases in some forests to compensate for the end of the old-growth in others, at the current average rate, all unprotected ancient forest will be gone by 2023.

Spokespersons for the timber industry often state that no more pro-tection of ancient forests is needed because so much is already "locked up." Although 18 percent of these national forests is wilderness, only 21 percent of their wilderness areas is ancient forest. To avoid incurring the timber industry's displeasure, most wilderness areas were carefully configured from scenic subalpine forest, alpine tundra, or rocks and ice rather than ancient forest. Thus, only 4 percent of the area of these six national forests is protected old-growth. Greene (1988) notes that 17,000 acres are protected in research natural areas on various federal lands. Assuming the 21 percent figure applies to the 197,000 acres of wilderness in the other three national forests, no more old-growth stands are protected and existing ones are not degraded, the entire inventory of ancient forest in 2023 will be about 1,178,000 acres (Table 8.6).

Table 8.5

OLD-GROWTH (O-G) IN WILDERNESS AND NONWILDERNESS (UNPROTECTED, AVAILABLE FOR LOGGING) IN
MORRISON'S SIX NATIONAL FORESTS
(ACRES IN THOUSANDS)

National Forest (total wilderness + unprotected)	Protected O-G in Wilderness	Unprotected O-G	Unfragmented, Unprotected O-G	Date All Unprotected O-G Gone	Date all Unprotected, Unfragmented O-G Gone
Mt. Baker-Snoqualmie (719 + 1,788)	119	178	116	2062	2009
Olympic (93 + 557)	16	90	66	2006	1994
Gifford Pinchot (173 + 1,081)	8	111	30	2008	1991
Mt. Hood (119 + 941)	44	134	57	2033	1996
Willamette (279 + 1,396)	105	194	61	2053	1997
Siskiyou (232 + 861)	60	80	18	2007	1990
Totals[a] (1,615 + 6,624 = 8,239)	354	784	348	2023[b]	

[a] Totals may differ slightly from sums due to rounding.
[b] Date all unprotected old-growth will be gone if the total cut from these six national forests does not change.

Table 8.6

PROTECTED ANCIENT FOREST IN WESTERN WASHINGTON AND OREGON
(ACRES IN THOUSANDS)

Where	Acreage
Wilderness areas: Morrison's six national forests	354
Wilderness areas: Siuslaw, Umpqua, and Rogue River national forests	41
Research natural areas	17
National parks	766
Total	1,178

Therefore, if current trends continue, in one generation, only six percent of the original old-growth will remain. Very little will be at low elevations. The actual numbers might vary slightly, depending on precisely how old-growth is defined and how the courts, the Congress, the federal land management agencies, and the American people view the elimination of the finest coniferous forests in the world. But the trend is unmistakable, and its implications are hardly encouraging. Nothing that conservation biologists know suggests that 94 percent habitat reduction bodes well for ancient forest species and ecosystem processes.

Sources of Bias in Old-Growth Estimation

□ *Elliott A. Norse*

Because key information was not available to Morrison, his methods have four sources of systematic bias, that is, bias that might consistently lead to under- or overestimation of how much old-growth remains. The first one can lead to underestimation; the other three can lead to overestimation. The four sources of bias follow.

1. *Succession.* Timber was inventoried on Morrison's six forests between 1974 and 1986. Since then, succession has probably caused some mature stands to become old-growth (as explained earlier, attainment of old-growth status is not simply a function of age), but it is not yet possible to estimate how much. Estimates of remaining old-growth at future dates are likely to be too low for the same reason.

2. *Downed logs.* The only criteria in the interim old-growth definition that could not be determined from timber inventory plots are the size, density, and biomass of downed logs. Exclusion of these criteria can lead

to overestimating the amount of remaining old-growth, but there is no way to assess the magnitude of overestimation from these data.

3. *Snags.* Timber inventory plots in Olympic National Forest had no usable information about snags, which can lead to overestimation. But this bias can be estimated by determining how many acres are rejected for insufficient snags on Morrison's other five national forests.

The portions of these five that had timber inventories would have had 1,156,000 acres of ancient forest if 357,000 acres (31 percent) had not failed to meet the minimum snag criteria. Thus, the best estimate for the actual acreage of old-growth in Olympic National Forest would be 31 percent less (33,000 fewer) than the 106,000 acres.

4. *Wilderness areas.* Neither amounts of logs nor any but the tallest snags could be determined in wilderness areas that had not been inventoried. This can lead to overestimation in those national forests (Mt. Baker-Snoqualmie, Mt. Hood, Willamette, and Siskiyou) where wilderness areas existed when the forests were inventoried.

Without an estimate of what percentage would fail the minimum old-growth log criterion, only the above 31 percent figure is available to deflate the 235,000 acres of preexisting wilderness that might have failed to meet the minimum old-growth snag criterion, although this is probably a bit high because at least some snags were visible. This accounts for some 73,000 acres.

Thus, the last three reasons suggest that estimates of remaining old-growth in Morrison's six national forests could be at least 33 + 76 = 109,000 acres too high without adjustment. The possible underestimation from (1) and overestimation from (2) tend to offset each other. None of Morrison's results is adjusted to account for these systematic biases, but the overestimation I can calculate would reduce by 9 percent the acreage of old-growth in Morrison's six national forests. □

Maintaining Our Options

Tis the part of a wise man to keep himself today for tomorrow, and not venture all his eggs in one basket.
(Miguel de Cervantes, *Don Quixote de la Mancha* [1605])

Microorganisms—bacteria, yeasts, and protozoans—are used as surrogates to study processes that also occur in plants and animals, includ-

Western redcedar logs, Olympic Peninsula (Washington). *These extremely valuable trees are being managed as a nonrenewable resource; they are not being replanted and do not reseed themselves quickly enough in tree plantations to provide marketable logs in the future. Sustainable forestry aims to protect biological diversity and ensure that forests can continue to yield valuable forest products such as these large redcedar logs.*

ing people. They are inexpensive to maintain, reproduce quickly, and evoke no outcry when you flush a billion or two down the sink.

If you take a bottle of sterile nutrient broth and seed it from a pure culture of protozoans such as *Tetrahymena*, the population grows rapidly for a while. Then it levels off and finally declines, perhaps to a low level, perhaps to extinction. Why? Because they have used up their food supply and generated toxic waste products.

Of course, this is an artificially simplified system where resources are nonrenewable and there are no other species to convert waste products into harmless or (better yet) useful substances. But *Tetrahymena* are normally part of a self-regulating system. If they overexploit their resources, their populations decrease, which allows other organisms—their food organisms—to recover. And their increasing wastes become a resource for other species that metabolize them and thereby render them safe for *Tetrahymena*.

Tetrahymena have survived because they live in a system—an *eco*system—in which their food supplies are indefinitely renewable and their waste products are indefinitely recyclable. The system is constantly adjusting to changes of all sorts with a combination of resistance, resilience, and adaptation. It is complex and messy, but it works. Such systems have been working nearly 2 million times longer than the time since Jesus gave the Sermon on the Mount.

We live in a system in which living things can provide for our needs for eons to come. *The* question facing the timber industry, forest managers, legislators, and the general public is whether to continue overexploiting and poisoning them, thereby crippling the life support systems that sustain us.

Consciously or unthinkingly, we are doing so. Our society is turning some of the world's most valuable renewable resources—Westside ancient forests—into nonrenewable ones. We are eliminating our finest life-support systems.

Some people might argue that the momentum is unstoppable. It's all going anyway, so we might as well squeeze the last profits from the system while we can. Others (presumably not young ones) might argue that it won't all be ruined in *their* lifetime; why not let future generations set things right? Still others say nothing because their bosses, customers, constituents, family members, or friends would not approve. If these beliefs prevail—and so far, they have—the ancient forests are lost. And they will take a lot with them as they fall.

But there is an alternative. It is not an easy one. It involves the willingness to rethink things we have always accepted, to learn from the evidence, to take prudent risks, to cooperate, to make trade-offs, to balance short-term and long-term benefits. The alternative is sustainable forestry, the new forestry.

Sustainable forestry means conserving the productive basis of the land by preserving the integrity of the biota and ecological processes and producing commodities without degrading the other values. It means forestry based on humility and appreciation, working with the land rather than against it. It means a sustained commitment to long-term research on the basic ecological processes that maintain values of both natural and unmanaged lands. It means clarifying and reinvigorating that much-abused term "multiple use" (or else abandoning it). It means viewing forests with a zoom lens, from microbes to landscapes. It means intelligently balancing competing interests so that neither those who would cut the whole forest nor those who would preserve every tree will get all they want. It means cooperation between old enemies to combat the real enemies of the forests. And it means making no irretrievable commitments that tie our hands as the world changes; more than anything, sustainable forestry means maintaining our options.

The current path threatens all we value from forests: species, jobs, ecosystem services, timber supplies, scientific understanding, scenic beauty, real estate values, and spiritual solace. But sustainable forestry can give us cathedral forests and peeler logs, good jobs and spotted owls, tourism and salmon runs.

I have no illusion that forging a consensus on sustainable forestry will be easy. The stakes are high, the issues complex. It will be a difficult, often frustrating, ongoing process. Everyone will have to make some concessions. But if we continue on our current path, it is clear that we will all lose, and lose big. At various times in this conflict, mill workers and Olympic salamanders, counties and ecological communities have all suffered major losses. Sustainable forestry is the "win-win" alternative to endless conflict.

The science and management know-how we need is not all there yet, although it will come if we give it enough time and resources. Yet despite all we don't know, the biggest hurdle is not inadequate scientific knowledge. It is the feelings of fear, mistrust, and hatred that have grown among participants in the conflict.

Loggers and environmentalists might seem to have little in common, but many share at least one thing: acceptance of the "either/or" fallacy: Either we have biological diversity or timber production.

It is a fallacy for two reasons. Some people look at forests as though they were simply multiples of trees, unconnected entities that can be measured with engineering precision and managed as if everything important about them were understood. But forests are far more complex than a nuclear power plant or a space shuttle. We don't understand the workings of any of the forest's components, and we understand their interactions even less.

Biological diversity is essential to sustaining timber production because we do not know what underlies forest productivity now, and know even less about it in the world of the future. Of course we know in a general way that trees need light, water, and nutrients. But even the best scientists cannot yet explain how forests work or why they sometimes do what we expect of them and sometimes do not. Most of what is known has been learned only since the start of two research programs: the International Biological Programme Coniferous Biome studies in the 1970s and the Old-Growth Forest Wildlife Habitats Program in the 1980s. To date, little of that knowledge has been used by the political leaders and managers who are making the decisions.

Without understanding what sustains natural forest productivity and stability, we cannot afford to eliminate the only forests that have withstood the test of time. What Aldo Leopold (1949) called the "cogs and wheels" of our finest forests is rapidly being discarded. Whether you care about curing cancer, reaching your timber quota, drinking some of the cleanest water left on Earth, bagging a buck, or reveling in hushed green serenity, you have to maintain biological diversity. You never know which of those cogs and wheels will turn out to be handy.

The converse point, that maintaining biological diversity depends on timber production, is not the specious idea that cutting trees increases biological diversity (see Chapter 4). Ancient forests did very nicely for millennia before settlers began deforesting the Northwest. But the Westside's exceptional value for growing trees provides a powerful incentive for maintaining existing forests and restoring former forests. And whether we want high-quality framing lumber or beautiful vistas, we face the same overwhelming threat to our interests, a common enemy: global environmental degradation, particularly atmospheric change. If anything can cause environmentalists and loggers to start talking, it is global change.

Until now, the Northwest has been polarized and paralyzed over the fate of the ancient forests. Few people have dared to offer ways to sustain all our forest values. Instead of candid discussions about our concerns, acrimonious lawsuits, media attacks, and lobbying campaigns have prevailed, forums in which extreme views have the upper hand. But in the long run, we cannot sustain either biological diversity or timber production unless there is widespread agreement that we need both. The only alternative to having neither is having both.

Prospects for Maintaining Nontimber Values

If we cannot end now our differences, at least we can help make the world safe for diversity.
(John F. Kennedy, speech at American University [1963])

The cut of old-growth on public lands is at record levels, yet timber industry employment is falling. Spotted owls and clear, knot-free lumber are both endangered species. A century of logging has eliminated all but 13 percent of the ancient forest in western Washington and Oregon. Six percent is protected; the other 7 percent, most of it in national forests and BLM lands, is what people are fighting over.

Have we already gone too far? What are the prospects for maintaining biological diversity and ecosystem services from scattered bits of ancient forest totaling one-eighth of what was originally there? or the one-sixteenth that will escape the chainsaw if current trends continue?

Even if no more ancient forest were cut and nothing else (such as climate) were to change, the prospects would be not be good. Everything conservation biologists know suggests that reducing habitat area by 87 or 94 percent leads to extinctions. Fragmentation only accelerates this process. Species with specialized habitats and large area requirements—such as spotted owls—are likely to go first. But they could have a lot of company, and most species at risk will probably be too elusive or too inconspicuous to be saved one by one. The only way to save them is to maintain their habitats.

Genetic diversity will decline even more than species diversity. Rare genes will be lost as populations are reduced and fragmented. Whole populations (with their distinctive genetic compositions) will vanish before species become extinct.

Ignoring pending changes in climate for the moment, the outlook for the diversity of remaining ecosystems might seem a little better. Both riparian ancient forests and natural stretches of large streams, for example, are already gone or nearly so. Scarcely any lowland old-growth remains in most of the region. But those larger stands that are left should maintain most of their species composition, structure, and functioning if they are allowed to. Perhaps a few very localized kinds of ecosystems will disappear due to changes already under way, but I would guess—guess is the appropriate word—that most remaining kinds of ecosystems could persist if no more were cut—*if* we could ignore changing climate.

In contrast, ecosystem services from ancient forests have been reduced at the same time that the rising population has come to depend more on them. Protection against erosion has probably been reduced less than might be expected from an 87 percent reduction because the first ancient forests to be liquidated were in the lowlands, where erosion is less severe than in the mountains. On the other hand, carbon storage has probably been eliminated to a greater extent because the densest stands with the biggest trees were logged first.

Fortunately, ancient forests are not the only ecosystems that harbor a diverse biota or provide ecosystem services. Earlier natural successional stages with substantial legacies of large living trees, snags, and logs from previous ancient forests can support many old-growth species, albeit in lower numbers. Even intensively managed tree plantations have *some* habitat value and can fulfill *some* of the ecosystem services of ancient forests. Consequently, the risk to ancient forest species is less than if most of their habitat were paved over (only some of it is). Ecosystem services have probably not been reduced 87 percent for the same reason.

We have far fewer options than we need to make sustainable forestry easy, but we are not bankrupt. There is still a chance.

The Two-Track Strategy for Sustainable Forestry

Let us give nature a chance; she knows her business better than we do.
(Michel de Montaigne, *Essays, Book III* [1595])

Given that too much old-growth has already been cut—that the highly reduced, highly fragmented remnants of ancient forest cannot sustain the biological diversity and ecosystem services we need—we have to be creative. Although ancient forests cannot be re-created, if we manage intelligently, a lot of biological diversity can be conserved on lands used for other purposes. This requires a two-track strategy for sustaining the special values of ancient forests: preservation of viable natural stands *and* sound management of other forestlands.

Preservation is essential because we do not know enough to manage for the incredible complexity that we do not yet understand. By most measures, the 6 percent already protected will prove inadequate to maintain ancient forest values. To avoid further losses, we have to know what we have, make the necessary decisions about what part of the remaining 7 percent needs protection, and know how to protect it. This would be difficult enough if global climate, UV-B levels and CO_2 concentrations were not changing; with these changes it will pose unprecedented challenges for those who design and manage forestlands.

As the widely different estimates of remaining old-growth forests show, there is an immediate need for a comprehensive inventory and mapping of forest ecosystems west of the Cascades so that we can make intelligent decisions on what to preserve, what to restore, and what to cut. For this to occur, some problems will have to be settled, for example, the lack of common definitions of old-growth among the twelve Westside national forests. We cannot afford to wait five or ten years to know what we've got. We desperately need answers now.

But the gaps in our knowledge go far beyond simple inventories. Our biological understanding of species that are not logged, hunted, or fished is woefully inadequate. We lack crucial information about even the best-studied nongame species—the spotted owl—and know far less about the others at risk. And knowing how they function individually is not enough; we need far more understanding about the functioning of communities, ecosystems, and the whole landscape, with special emphasis on how human alteration affects species and ecological processes. All the research thus far has provided only part of the foundation needed to deal intelligently with the profound landscape and atmospheric changes that lie ahead.

And we need to begin deciding what must be preserved so that we can maintain our options. Even if we knew what we had and how it works, deciding how much to save, how much to restore, and how much to liquidate will be painful. At present, decisions are being made appropriations bill by appropriations bill, timber sale by timber sale, court battle by court battle. One side or the other wins or loses (for the moment), but public land-management agencies have no plans for managing in ways that will not rapidly liquidate the ancient forests. Armed with the needed facts, recognizing the need to balance competing interests, we must have a better way of making decisions while we still have options.

To manage lands for biological diversity, we need to invent a new forestry, a sustainable forestry that goes beyond the forest farming to which most modern silviculture aspires. Growing trees that can live a millennium is very different from growing annuals. And trying to make them into annuals will not work, either ecologically or economically. The Westside's special advantage as a timber-producing region is in production of high-quality wood. Why maximize volume and ignore quality? Why try to play the game that the South—with higher growth rates, flatter lands, closer proximity to markets, and lower labor costs— will certainly win? Why not play the game for which our forests are best suited? To do it, we have to take a longer view. And that means maintaining biological diversity.

Practicing sustainable forestry will require all the skills that silviculturists can muster. Until now, forest management has been synonymous with timber management. Land-management agencies have long talked about values such as wildlife (game species that forage in clearcuts), recreation (in motor vehicles that use logging roads), watershed management (by clearcutting), and beauty (by cutting down the green obstructions to vistas). But nontimber values (when mentioned at all) have been used mainly to justify yet more roading and logging. President Theodore Roosevelt didn't choose to establish a "Timber Service" for good reason. America's public forests are more than timber.

Instead of liquidating our forest capital, we need to learn to live off the interest, making no renewable resources nonrenewable and sustaining production of diverse goods and services. And we must produce them in a world that is rapidly changing.

This would be a major shift. Because the monetary value of wood products is large and easily measured, timber considerations have dominated decisions about our forests. But most values from ancient forests are not market values. It is harder for economists to quantify the benefits of pristine air, superb drinking water, quiet, wildflowers, songbirds, catching a steelhead, ameliorating flooding, or preventing global climatic change. The beneficiaries of tourism are spread throughout the business community; salmon fishermen are not organized to protect their interests as well as the timber industry. And the harvest of nontimber commodities, such as wild mushroom and cascara bark that are worth millions of dollars annually, is largely ignored by agencies that focus on maximizing the volume of timber. For that matter, so is clear, strong heartwood.

To maintain our options, we need a forestry that recognizes that our needs and values are diverse, one that allows us the flexibility to respond to change. A century ago, western white pine was probably the most desired forest product. Had anyone been planning for the future then as we are now, tree plantations might have had a lot more pines. Of course, no one could have anticipated that white pine blister rust would eliminate 95 percent of them.

Today we are putting all our eggs in a different basket: planting huge acreages with Douglas-fir monocultures. They could be at least as vulnerable to future changes as white pines were. We face not only rapidly changing regimes of insects, diseases, fires, and climate, but also changes in demand. The most desired forest product in fifty years may not be young Douglas-fir wood; it may be redcedar or alder wood or yew bark; it may be the chance to collect chanterelles, catch wild cutthroats, photograph spotted owls, drink from a stream, or just enjoy the beauty of nature. Just as an intelligent investor diversifies a portfolio as a hedge against uncertainty, an intelligent silviculturist maintains biological diversity in managing for a future that is certain to be uncertain.

Preservation and active management complement each other. Preserving the range of ancient forest ecosystems provides managers with irreplaceable models of how forests can survive and function over long time spans in the face of change, provides the species and gene resources needed for the future, and serves as a hedge if our managed forests don't turn out the way we wanted. They can serve as a buffer against change; for example, with their much greater diversity of predators, ancient forests may be effective barriers to pest outbreaks sweeping through tree plantations. And ancient forest preserves offer the untamed recreational experience that could provide the greatest economic returns from our forests, as Randal O'Toole (1988) points out. Wilderness is already in shorter supply than wood, and the gap between them will only widen.

Silviculture, in turn, can benefit ancient forest preservation. Long-rotation management should be able to provide high-quality wood that would remove pressure from ancient forests. Further, managing surrounding landscapes to minimize differences with ancient forest remnants would help to maintain the viability of the fragments. And silviculture can be used to increase population sizes and ranges of ancient forest species in managed forestland. Many benefits will arise from collaboration among silviculturists, landscape ecologists, forest entomologists, and conservation, wildlife, and fisheries biologists.

The intellectual challenge of the new forestry is great, the risks are real, and the penalties for failing are large. But the alternative—business as usual—is already failing and will continue to fail to meet the changing demands of a changing world.

The most difficult obstacle to developing a new forestry is not a lack of scientific understanding, although we certainly need better understanding for the task ahead. It is not a lack of silvicultural skills, although forestry schools need to reshape their curricula to reflect the needs of the future. Rather, it is human nature: the tendency to see things as "either/or," to base decisions on expediency, unquestioningly to accept old paradigms, and to fight against, rather than cooperating with, nature and one another.

Congress is still placing forest managers in the untenable position of cutting more timber than forests can sustainably produce. Management agencies are still reserving the fast career track for men who meet timber quotas. Universities are still producing students who measure forestry goals in board feet. The most productive federal forestlands are still managed on one track, a one-way trip to somewhere the American people—wisely—do not want to go.

To have a forestry suited to a changing world, these must change. The hurdles on the path are less scientific, technological, or economic than political, sociological, and, ultimately, psychological. But I am willing to wager that our foresters are up to the challenge.

Preserving Ancient Forests Is Not Enough

Nothing endures but change.
(Heraclitus [ca. 540 to ca. 480 B.C.E.], from *Lives of Eminent Philosophers, Book IX*, by Diogenes Laertius)

All ecosystems are ephemeral. Even in the depths of the oceans (the most constant places in the biosphere) scientists have observed powerful undersea storms devastating the ancient communities of seabed dwellers. On land as in the sea, change can be sudden or gradual, global or local, periodic or random, frequent or rare.

A 1962 storm blew down 26 million cubic meters of timber. The 1980 eruption of Mt. Saint Helens destroyed 128,000 acres of forest. Greater eruptions could come from Cascades volcanoes at any time; the eruption of Mt. Mazama 6,600 years ago (creating Crater Lake) ejected forty times more material. During 1987, fires swept through some 800,000 acres in southwest Oregon and northwest California. More common are local windstorms, ice storms, floods, landslides, and outbreaks of insects and pathogens. To these we are adding new atmospheric changes.

Although Douglas-firs need disturbance to establish, disturbances can kill them as well. In Washington, disturbances are frequent enough that very few reach their 1,200-year potential life spans. Few stands survive to be succeeded by western hemlocks or silver firs. In southwest Oregon and northwest California, where fires are more frequent, few Douglas-firs reach 500 years. Most old trees bear fire scars around their bases. Survivors are either resistant or fortunate.

In time, disturbance will eliminate today's old-growth stands. Hence, preserving ancient forests is an essential element of any conservation strategy, but it is not sufficient. Rather, it is necessary to preserve the processes that yield ancient forests by protecting a full range of natural successional stages. Last year's burns or this year's mature forests will become tomorrow's old-growth.

How much natural forest must be protected to ensure a future supply of ancient forest? The answer depends on the kind of old-growth that we are aiming to have.

For example, the amount will depend on the rate at which stands

Table 8.7

HYPOTHESIZED AMOUNT OF LAND NEEDED TO YIELD 100 ACRES OF
500-YEAR-OLD FOREST

Stage (years)	Mean Stand Mortality/ Year (%)	Likelihood of Stand Surviving (%)	Acres at End	Acres at Start
Old-growth (225–499)	0.10	73	100	138
Mature (80–224)	0.20	71	138	194
Young (25–79)	0.50	73	194	268
Sapling (5–24)	0.50	90	268	298
Seedling (0–4)	2.00	90	298	331

develop old-growth characteristics (the best sites probably develop them fastest). But as a general rule, the older the old-growth we want, the more younger forest will be needed to get it because disturbances can eliminate stands throughout their development.

Another factor to consider is vulnerability to disturbance. Some sites are visited by fires, damaging winds, landslides, floods, insect outbreaks, or even damage by volcanic eruptions more often than others. For example, the oldest stands are—or at least were—almost always in moist riparian zones, which inhibit fires. Further, human alteration of the landscape and atmosphere alters the vulnerability of particular sites to disturbance. Adjacent fire-prone young tree plantations or clearcuts that increase wind fetch increase the vulnerability of a future old-growth stand. In general, the greater the risk of disturbance, the more young forest will be needed to yield a given amount of old-growth.

We can construct a crude model that makes some oversimplified assumptions about forest community dynamics and does not account for the increasing vulnerability to disturbance due to atmospheric and landscape changes. The numbers are pure guesses and are undoubtedly off the mark. But the approach should be useful nonetheless.

To plan for a 100-acre stand of a 500-year-old Douglas-fir forest beginning with a burn that is immediately colonized by Douglas-fir seedlings, we might break down the stages as shown in Table 8.7. It predicts that an average of 3.3 acres of land is needed to yield 1 acre of 500-year-old trees. Of course, the number is very sensitive to mean stand mortality/year. If the figures used are too high or if the forest we seek is not such old old-growth, fewer acres will be needed to begin. (Especially in view of global atmospheric changes, it is also possible that mortality figures are too low.) But the basic finding should hold: To get 100 acres of old-growth requires a considerably larger amount to allow for the inevitable loss of stands.

Maintaining Biological Diversity in Managed Forests

Nature, to be commanded, must be obeyed.
(Francis Bacon, Novum Organum [1620])

Silviculture provides powerful tools for manipulating forests. But silviculture is no better than its goals; if the only goal is to maximize short-term wood production, other values will inevitably suffer. Similarly, forest management has been constrained by its reliance on traditional silvicultural tools. I have never heard a more insightful statement of this than in the saying, "If the only tool you have is a hammer, you tend to treat everything as if it were a nail."

The new forestry uses both traditional silvicultural tools and new ones to sustain biological diversity, ecosystem services, recreation, scenic beauty, and timber production.

Some objectives will rarely conflict. A forest managed for biological diversity can provide high levels of ecosystem services and aesthetics. But the more objectives a forest manager has, the greater the likelihood of conflicts and the more professional ingenuity he or she will need to resolve them.

Some foresters might not feel up to the task; in the short term, it might seem easier to farm trees like corn. But the easiest way is not always the best way. Fulfilling multiple objectives on a sustainable basis is a greater challenge with greater rewards than simplifying forest ecosystems to fulfill a single purpose. It should not intimidate the best and brightest foresters. Managing forest ecosystems (rather than just timber) requires mastery of a broader range of tools to suit the individual needs and limits of each site. Fortunately, many already exist, and others, such as powerful but user-friendly geographic information systems, will undoubtedly be developed as the need becomes apparent.

Silviculturists, wildlife biologists, fisheries biologists, hydrologists, conservation biologists, ecologists, and people with similar professional skills, working together, should be able to develop creative ways to manage forest ecosystems for multiple objectives. The following sampling of ways to maintain biological diversity of managed forestlands will require trade-offs. Some could lessen short-term yields of timber to varying degrees. But they will also have many benefits in terms of ecosystem services, recreation, aesthetics, and long-term site productivity.

1. *Manage on spatial scales appropriate to all your objectives.* Most public land is managed on the scale of a timber scale (10–25 ha). This ignores disturbance processes that occur on both smaller (individual treefall) and larger (watershed to landscape) scales. Patch shapes on managed lands are quite regular, even if their edges are rounded. The result is a regular mosaic of homogeneous patches, a pattern very unlike the mosaic of uneven-sized and -shaped, heterogeneous patches produced by natural disturbances.

Leave trees, Gifford Pinchot National Forest (Washington). *Forest managers can increase biological diversity and speed forest recovery by mimicking natural forest processes, for example, by leaving scattered trees or, better yet, clumps of trees in logged areas. This does not substitute for preserving ancient forests. But trees that are carefully chosen to withstand wind can provide needed perches and nesting places during the decades when they are otherwise absent in tree plantations.*

This spatial simplification diminishes biological diversity because the resource-gathering, mating, and dispersal strategies of many organisms are tied to particular scales of disturbance that do not occur in managed forest landscapes. It also encourages outbreaks of a few species—often ones we consider pests—by creating ideal conditions for their dispersal. Ecosystem management consciously mimics natural disturbance patterns that different organisms need by finding appropriate mixes of preservation and by cutting on at least a watershed (or better still, on a ranger district or forest) scale.

When clearcuts are used, they should vary much more in size and shape than under the staggered-setting system. To minimize fragmentation and save on roading costs, they should be aggregated to increase effective patch size.

2. *Make clearcuts "messy."* Virtually no natural disturbance creates large patches as bare as billiard tables. This is clear even at the site of a

holocaustic fire; many nonliving structural elements and organisms remain as legacies vital to regeneration. Where clearcutting is preferable from all the standpoints of sustainable forestry, managers can still maintain biological diversity by resisting the pressures from our parents and traditional forestry teachers: "Clean your plate—children in Europe are starving!" "Fix your bed!" "Cut your snags—they could fall on someone!" and "Burn your slash," and so on. Minimizing biological diversity on your plate is one thing, but doing so in your forests is another. Excessive neatness needlessly eliminates the microhabitats that many species need.

The larger a clearcut is, the more important it is to cut in an irregular shape with feathered edges, to leave remnant patches of various sizes within the cut, and to leave adequate numbers of snags and logs so that at least the more disturbance-tolerant forest species have a chance of surviving until surrounding conditions become suitable again. When the urge to tidy up those cull logs becomes overwhelming, just imagine your mother saying, "Put those back where they belong (voles and salmon fingerlings need habitat too!)."

3. *Manage for linear features as well as for patches.* Linear features, including burns, avalanche chutes, and riparian corridors, can be important landscape elements. Wildlife biologists have long known that forest edges benefit some species. This has been seized upon to justify accelerated roading and clearcutting. Unfortunately, edges also serve as both pathways by which edge species enter the forest interior and as snowy barriers to movement in winter.

It is less widely known that forested riparian corridors can serve as snow-free highways for terrestrial animals moving up- and downhill. The practice of leaving narrow riparian buffer strips should be extended to include small headwater streams, and buffer strips should be designed with migration in mind. Current buffer strips broad enough to provide shade are seldom broad enough to lessen snow depths. Even strips of wind-firm trees that cannot provide forest interior microclimates can still offer cover, perches, and food for some dispersing ancient forest species, and they merit wider use. If it is impossible to provide continuous corridors, linear archipelagoes of remnant patches might have value for some of the more mobile species.

4. *Manage for uncommon ecosystems.* Streams, seeps, and swamps make a disproportionate per acre contribution to biological diversity. Although it is wise to leave some remnant patches in high spots (it is

easier for organisms to disperse downward than upward), it is especially important to leave remnant forest around swamps and seeps. The same is true for any rare, highly dispersed ecosystems.

5. *In managing succession, take your cues from nature.* All the vegetation management studies I have seen have measured conifer growth over the short term—a few years—rather than throughout one or more rotations. This makes the assumption that the only effect of noncommercial early successional plant species on conifers is competition. While they do need the same water and light that conifers do, and can slow growth during early succession, they can also leave a large pulse of nitrogen that eventually moves through the trophic pathways of the whole ecosystem, a benefit whose results might not be fully visible for many decades. Nitrogen demand is decreased and N supply is increased (as it is released by decomposition) in the first years after clearcutting, so N is less likely to be limiting than in later years. Because succession usually ensures that "the first ones now will later be last" (as Bob Dylan observed), it makes sense to hedge our bets and go easy on early successional species.

Furthermore, our knowledge of species interactions within the soil is so limited that we could well find that "competing" species play roles no less important than nitrogen fixing, such as disease suppression, serving as hosts to mycorrhizae that are needed by later successional species, pumping essential elements to the conifer rooting zone from below, and restoring channels in the soil.

Until we can verify that these competing species are so "bad" *throughout a rotation* that they must be suppressed, resisting the urge to spray will benefit everything from bees to elk. A halfway step between letting nature take its course and killing everything but Douglas-firs is practicing selective vegetation management after other plant species have contributed their benefits to the ecosystem for a few years. Wherever possible, manage by nudging, not by bludgeoning.

6. *Allow (and encourage) natural regeneration on some sites.* The diverse plants that recolonize after a disturbance are not only important components of biological diversity in themselves; they provide essential resources for many other species. The diversity of species and ages, and the uneven canopy that develops, could be important hedges against the pests and crown fires that might be more frequent visitors to forests in our warmer future.

7. *Plant a variety of species appropriate to the site, if not interspersed, then in patches.* Monoculture of even-aged trees unquestionably saves

certain costs, but it maximizes risk for timber production, and it is the worst possible silvicultural system for biological diversity because it offers minimal spatial and food resources. Intensively managed even-age polycultures could be an alternative to monocultures that would offer more habitats and food resources for animals and decomposers. Even planting interspersed blocks or strips of different species, such as Douglas-firs, red alders, and western redcedars, could provide more kinds of opportunities than Douglas-fir monocultures.

These principles are only a beginning; they do not replace the preservation of large, viable natural forests. But they could serve as essential complements to preservation where there is anything left to preserve; in other places, they offer the only means of increasing biological diversity beyond the little in an intensively managed tree plantation.

Attaining Environmental Maturity

But ask now the beasts, and they shall teach thee; and the fowls of the air, and they shall tell thee: Or speak to the earth, and it shall teach thee. . . . With the ancient is wisdom.
(Book of Job, The Bible, King James Version [1611])

The ancient forest debate has all the elements of environmental controversies in general. One group asserts its right to alter something that the other group values and wishes to protect. The fact that this will deplete or eliminate the resource being contested, harming everyone's interests, seldom deters those who are doing it.

In trying to understand why people would do this, I have to cross the usual boundaries of science and public policy analysis and look to an insight planted in me when I was about ten years old.

My mother told me that when I was a small child, I thought she was all-knowing. She predicted that when I became a teenager, I would find her (and all adults) incredibly stupid. Finally, she said, when I became an adult, I would come to see that she wasn't quite so stupid after all.

Of course, she wasn't suggesting that her intelligence would change; what would change were my feelings, from awe to contempt to appreciation.

There are three comparable stages in our relationship to another "mother," the one we call nature. As tribal societies, we were awed by her incomprehensible power. We offered sacrifices to her forces and called them gods of rain and wind, sun and fire.

As our power grew, we began to defy nature, eliminating the vast herds of game, turning Mediterranean forestlands into semideserts, and fouling the water, atmosphere, and land. Such defiance is still widespread. Just as teenagers disdain parents they once respected, we disdain nature. Like them, we paint the world in simplistic terms: We can have either jobs or old-growth forests, not both.

The analogy goes further for people on both sides of the debate. There are those at logging rallies who gun the engines of their truck hot rods and distribute bumper stickers saying, "Sierra Club: Kiss My Axe" and "Save a Logger—Kill a Spotted Owl." And then there are environmentalists who spike trees. These are not adult responses.

Adults learn to see the world in its complexity, to live as well as possible within their means, to delay gratification, to see the value of laws, to use their power responsibly, to balance, to compromise.

Since Euro-Americans replaced Native Americans, the environmental ethics with which we have treated Westside forests have been largely those of the second stage. But as ancient forests shrink worldwide and the consequences of recklessness become harder to ignore, we are entering the third stage. I believe that it is the stage in which Northwesterners will protect and cherish what remains of our forests, yet use them for the economic bounty they can provide in perpetuity. It is the stage in which silviculturists and ecologists will argue not about intensification versus preservation, but how best to use our professional ingenuity to perpetuate forest values. It is the stage when talk of balancing the interests of different users will cease to be hypocrisy. It is the stage of understanding and appreciation.

Just as people do not mature overnight, the transition to environmental responsibility is far from complete. But despite many discouraging signs, I am convinced that it has begun. As we learn to appreciate the rich living systems that are unique to the Westside, we are entering environmental adulthood.

IX
Conclusions and Recommendations

Conclusions

. . . when nature is thought of as evil, you don't put yourself in accord with it, you control it, or try to, and hence the tension, the anxiety, the cutting down of forests, the annihilation of native people.
(Joseph Campbell, *The Power of Myth* [1988])

Many Northwesterners think of the ancient forest issue as a local one. But *all* environmental issues are local. Some, however, have far-reaching, even global, significance. Loss of Westside ancient forests is one. Standing back, taking the broad view, it is not hard to see why.

From our origins on the African savannas, our species has covered the world. No other animal or plant occupies so many habitats. Our unique ability to change our surroundings has allowed us to achieve success unmatched in the 3,500-million-year history of life on Earth. Yet this behavior, as characteristically human as our large brain, upright posture, tool making, language, and sociality, is proving to be our undoing. We are destroying our planet, our only home.

When our population was small and our tools were made of stone, it mattered little whether we hated or loved nature. Living systems were resilient enough to absorb our insults. We were just another distur-

271

bance, on a par with a hurricane or locust outbreak. But now the 5,200 million of us (as Peter Vitousek of Stanford University and his coauthors [1986] have shown) appropriate almost 40 percent of the land's net productivity, and we are a disturbance outweighing all the storms, fires, and plagues combined. As when the Earth experienced its last great cataclysm 65 million years ago, apparently from collision with an asteroid, we are experiencing a mass die-off of living things. But many of the world's most eminent scientists believe that this mass extinction—one we are causing—will be even more damaging.

The enormity of this act has no historic equal. But even putting aside any moral considerations and simply looking to our own self-interest, it does not make sense to destroy the life of our planet.

What can we do?

There are two components to any answer.

The first is population. There are more people than the Earth can support without being degraded. Overpopulation is not just in Calcutta or Los Angeles, Seattle or Portland; its signs are unmistakable for anyone who knows what to look for. Just ask yourself: Can I go to the closest body of fresh water and drink from it? Can I always see the mountains when it's sunny? Do I often spot elk, bears, beavers, or eagles when I take a walk? The people who lived in the Northwest for millennia would have answered yes and wondered why we would ask such foolish questions. If we do not find an answer to overpopulation, it will answer itself in very unpleasant ways.

The second is what we do, as groups and as individuals. Americans' resource consumption exceeds what the Earth could sustainably support even if we were the world's only people. But we are only 5 percent of the world's people, and the rest want to live exactly as they see us living on *Dallas* and *Dynasty*.

What would happen if they achieved our level of resource consumption? The landmark December 1988 issue of *National Geographic* points out that the United States has 135 million registered cars. India, with over three times our population, has only 1.4 million, but if Indians lived the way we do, they would have 440 million cars. That would undoubtedly gladden General Motors and Toyota, but how would it affect global energy consumption and atmospheric pollution? Our attitudes and appetites set an example for the whole world. People in other nations might not emulate what is best about the United States, but they certainly want to emulate what they believe to be our lifestyles.

What does this have to do with the Westside? For one thing, we have the good fortune to have the world's finest coniferous forests within our borders. They are extraordinary for their cathedralesque beauty, for their production of fine timber, for their rich diversity of living things, for providing abundant water of astounding purity, for storing more carbon per acre than land ecosystems anywhere in the world (several times more than tropical forests).

Brazil and the other nations of the Amazon basin also have extraordinarily valuable ancient forests. But it is becoming abundantly clear that Brazil does not appreciate hearing that other nations want it to protect

Rob Bierregaard

Former ancient forest (state of Amazonas, Brazil). *Third World nations desperate to improve living standards and to pay off international debts are rapidly destroying their ancient forests. As much as 15 percent of Brazil's Amazonian forest is already gone. The United States is encouraging tropical nations to stop destroying their forests. Destruction of our own ancient forests is far more advanced—87 percent of Westside ancient forests have already disappeared.*

its forests. It wants jobs for its 145 million people. It wants to repeat our pattern of development, to provide its people the same affluence we have, to become a world power to be reckoned with. And Brazil isn't alone in wanting these things. Virtually every developing nation does.

For better and worse, our nation no longer has the raw power to coerce developing nations to do what we believe is best for them, for us, and for all humankind. But it is within our power to do one crucial thing: We can set a shining example. We can take care of our own ancient forests. Conservation begins at home.

The Westside faces many of the same problems that people face worldwide; indeed, rural Oregon's excessive dependence on a single, fast-disappearing commodity reminds many observers of a Third World nation. But the Westside is in a vastly better position to preserve and sustainably manage its forests for two reasons.

The first is its land. The Westside is uniquely suited to great forests: Its soils are rich, its climate is ideal, and its dominant ecosystem is highly productive and resistant to environmental changes. Its biological diversity provides outstanding economic benefits. And it is exceptionally safe from the air pollution that increasingly plagues forests in other developed regions.

The second is its people. The Westside's population growth rate is more manageable than that of India or Brazil, Florida or southern California. Standards of living are high enough that few people are driven to desperate acts because their basic needs are unmet. Most people are well educated, an invaluable asset in the quest to find far-sighted, innovative solutions to problems. The region's legislators and officials are known for their honesty and sensitivity to what they perceive to be the will of the people. Most important of all is the prevailing mind-set; as a group, Westside people are exceptionally attuned to the land and (with the exception of this issue) are the most environmentally conscious people anywhere in the United States. The people of the Westside have found solutions to environmental problems that have thus far proven insoluble almost everywhere else. If any region can sustain its primary resource through future changes, it is the Westside.

But so far it has not (Figure 9.1). In the century it has taken the population of Washington and Oregon to increase ten times, nearly nine-tenths of their Westside ancient forests have disappeared. Only one-sixteenth is now protected, and 3 percent or more of what is not protected is now being destroyed each year, so it will all be gone in just thirty-four years. We are destroying our source of wealth and well-being

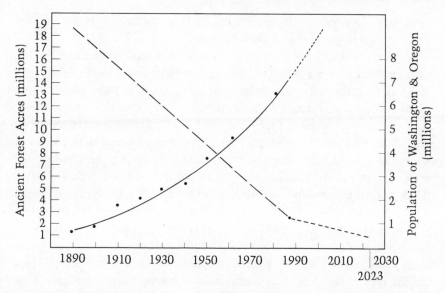

Figure 9–1. **Population growth in Washington and Oregon and loss of ancient forests in western Washington and Oregon.** *Population has risen by about a factor of ten since 1890, while ancient forests have declined by almost a factor of ten. If current trends continue, only 6 percent of the ancient forests will remain in 2023.*

no less than Ecuadorians, Ugandans, and Nepalis, who have far fewer resources to deal with their far greater problems.

The Westside economy was built on a timber industry completely dependent on consuming an irreplaceable resource that nonetheless seemed both endless and valueless except as timber. Now we know that our living resources are more valuable and are being exhausted faster than anyone had imagined. We are in urgent need of solutions, and we need them not in ten years, not in five years, but now.

Offering solutions to overpopulation and wasteful end-use of resources is beyond the scope of this book. But the way that we treat the land is not. It has changed and is continuing to change.

When white men settled the most thickly forested land on Earth, they considered it an enemy, mowing down the forest with no thought but to "cut and get out." Gradually, however, they came to treat the land more as a valued but unruly child that had to be disciplined until it behaved according to their expectations. This transition parallels the ways they treated people of other nations, races, sexes, and ages. Such paternalistic attitudes have now relaxed considerably in many areas of life, but forestry has lagged behind most others.

Economics and laws now prevent loggers from cutting and getting out. But the current system of replacing ancient forests with tree plantations is not much better; indeed, in some ways it is worse. It short-changes everyone—from today's sightseers to tomorrow's loggers—by foreclosing our options. Intensive timber operations—extensive roading, short-rotation clearcutting, slash burning, planting with monocultures of genetically impoverished trees, spraying herbicides, and suppressing natural disturbances—do not replace what they destroy. They take a miraculously complex, indefinitely self-sustaining forest ecosystem and convert it into a long-lived coniferous cornfield.

Perhaps it can be argued that forest farming is appropriate for private lands, but it is not for public lands. Private lands are more productive, so trees grow faster. They are flatter, so soil erosion is less a concern. These advantages make them more amenable to intensive, short-rotation forestry than public lands. And in fairness, timber companies have been no harder on the biological diversity of lands they own than have farmers or commercial developers. But governments do not allow farmers and developers to do what they do on lands owned by the public. Their treatment of loggers is quite different.

On private and public land alike, eliminating biological diversity eliminates the natural regulatory mechanisms that have sustained forest ecosystems through millennia of storms, fires, and insect outbreaks. The densely stocked young tree plantation monocultures that are covering the Westside are inherently unstable, so foresters feel compelled to manage more and more intensively. Unfortunately, the continued manipulation needed to maintain tree plantations is not only increasingly expensive and environmentally damaging; there is a substantial risk that it will fail to deliver the timber in the face of landscape, climatic, and economic changes. The seedlings being planted today will approach rotation age in a very different world.

Yet, at a time when changing markets, increasing temperatures, ultraviolet-B radiation, and atmospheric carbon dioxide are altering conditions in ways that we cannot predict, managers are "putting all their eggs in one basket," jeopardizing both the remnants of natural forests and the sea of tree plantations surrounding them. Strategic planners in other fields—retail sales, finance, energy, agriculture, defense—know that the first rule in times of uncertainty is to maintain a diversity of options. Foresters need to learn this too.

Fundamental changes are needed. Jack Thomas and his coauthors summarize why:

> Existing knowledge and professional judgment in the area of forest ecology are too often underemphasized in setting policy and planning for management of publicly owned forestland. Objectives for wood fiber production often are based largely on considerations such as historic levels of timber harvest, capacity of mills, and consideration of economic factors. While such decisions may be economically compelling, they are ecologically naive. Wildlife and other ecological values are too frequently considered by some only as "constraints," but are, in reality, important values (and are defined as such in law) that should be embodied in the management objectives initially. Wildlife should be viewed as one of several joint products from managed forests, not a constraint on timber production. (1988, p. 257)

We need to return to the principles that are supposed to govern federal land management agencies: that our land resources belong to *all* Americans, not just those who get their money from logging. We need to redress the imbalance between immediate gratification and long-term benefits, to shift from maximizing the cut in the short term to sustaining biological diversity and timber production far beyond the next cut. We need a dramatic increase in the role of science in managing public forestlands. Because we face very real threats to our forests from beyond the region, we need to foster cooperation among various interests to devise creative ways to deal with them.

Rather than whipping our forestlands into submission, we need—to borrow a phrase—a kinder, gentler—and wiser—forestry.

No one is blind to the need for good jobs for the thousands of people in the timber industry. But hurrying the destruction of the resource that has sustained the industry can only hasten mill closings. The ancient Greek storyteller Aesop had a name for it: killing the goose that lays the golden eggs.

To avoid this, I offer some broad recommendations for changes that we need to make to sustain our forests. It is not my intention to suggest many specifics; that is something best done by bringing all the interested parties together. I believe that "victories" in court or in Congress will be temporary unless there is substantial movement toward consensus among those with a major interest in the future of our forests. These recommendations provide all of us a chance to "win."

Those who prefer fighting to winning will undoubtedly dismiss them as unworkable. But those who can see that we are headed for unpleasant collisions should see in them the means of avoiding preventable loss of our great forest wealth. If I have succeeded in helping readers to understand what is at stake and how it works, your enlightened self-interest should ensure that we succeed.

Recommendations

. . . the scientist is wiser not to withhold a single finding or a single conjecture from publicity.
(Johann Wolfgang von Goethe, *Essay on Experimentation* [ca. 1830])

These recommendations are based on one principle: Because biological diversity is the source of *all* we value from Westside forests, maintaining it is the *highest* priority for forestland management. Biological diversity is the key to sustainability. Whatever other goals we have, the first and foremost among them must be preventing the loss of genes, species, and ecosystems. In the Westside, the primary way to do this is by stopping the destruction, fragmentation, and simplification of ancient forests, especially the rarest types, including riparian and other lowland forests.

This does not mean that consumptive uses of forests have to stop. Logging, gathering plants and fungi, and hunting and fishing are important activities that need not degrade forest values if we understand forest ecosystems well enough to manage them intelligently. But our first priority must be ensuring that we do not destroy the very source of all their products and services.

The primary job of forestland management, then, is ensuring that human activities are fully compatible with maintaining biological diversity. Successful managers will preserve the natural mechanisms that have sustained the productivity and integrity of the forest for thousands of years, mechanisms essential for coping with future changes. They will sustain not only wildlife, clean water, and beauty but timber production as well. Once this goal is understood and accepted as paramount, the changes needed for successful forest management become much clearer, and the current conflict will diminish markedly.

In keeping with this goal, I offer the following seven recommendations.

INVENTORY AND INTERREGNUM

It is impossible to have a sound strategy to manage all the values of public forestlands unless we know what is there. It is essential that the Forest Service and the Bureau of Land Management inventory the forests and cutover lands on all public and private lands throughout the Westside. The inventory should assess the locations, amounts, and spatial relationships of natural and planted stands of all age classes and species compositions, and place these data within a single, user-friendly geographic information system. This inventory will provide agency employees, conservationists, communities, and the timber industry with the information necessary for decision making. It is of highest importance that the agencies use uniform, agreed-upon definitions, methods, and quality assurance techniques so that users of the data can get a meaningful, comprehensive picture of the Westside's forest resources. Indeed, the agencies should confer with these and other interested parties in developing definitions and methods of displaying data.

The loss of ancient forests has reached crisis proportions; "business-as-usual" is destroying them. The American people cannot wait five or ten years to know what we have. To minimize both forest loss and economic dislocations caused by further delay, the agencies should allot adequate staffing and funding to expedite the inventory.

Once the first stage of the inventory is completed, the status of all public lands should be updated on a continuing basis against the records of all decisions and actions of land management agencies, and, independent of this updating, the inventory can then be updated by satellite or aerial surveys and ground truthing at least every five years.

Until the inventory is finished, public land management agencies need temporarily to stop both timber sales and logging of existing sales of any stand that meets the definitions given by the Old-Growth Definition Task Group in 1986. The purpose of this interregnum is to prevent further loss of options until decisions can be made on how much old-growth must be preserved and how the rest must be managed. Until the inventory is completed, the interregnum will decrease logging from the high levels of recent years. However, the management agencies already possess substantial amounts of mature forest, some of which can be scheduled for logging to minimize the adverse economic impacts on people who now depend on cutting old-growth on public lands. At the same time, the federal government and state governments can work with business and community leaders to diversify local economies.

PRESERVATION

With the inventory in hand, land management agencies throughout Westside California, Oregon, and Washington can use its results to develop comprehensive plans with two purposes: (1) preserving the lands judged by the best available scientific opinion to be necessary for long-term maintenance of forest values, and (2) managing for long-term maintenance of forest values on other lands on which logging is allowed. These plans would form the basis for decisions to preserve lands as national parks, wilderness areas, or, as proposed by The Wilderness Society's David Wilcove (1988), diversity protection areas. Congress, the public land agencies, and other authorities should ensure that these plans come to fruition, and each BLM district and national forest should be held accountable for implementing them.

The Northwest has perhaps the most serious gap in the system of preserved lands in the continental United States: the Klamath-Siskiyou region, biologically one of the richest areas of North America, indeed, one of the richest temperate areas in the world. Like the Great Smoky Mountains, the Klamath and Siskiyou mountains are a geologically complex zone of overlap between North America's northern and southern forests. And like the Smokies, the extraordinary biological diversity of the Siskiyous made them a primary study area where renowned forest ecologist Robert Whittaker (1960) of Cornell University worked to discover basic principles of community ecology. More than fifty years ago, our nation made a crucial step in protecting our biological diversity when President Franklin D. Roosevelt designated Great Smoky Mountains National Park. Its western analogue, the Klamath-Siskiyou area, warrants the same protection.

With the full cooperation of the Forest Service and the BLM, the National Park Service needs to prepare a new area study to establish a national park within the Klamath-Siskiyou area. Then Congress should designate a Klamath-Siskiyou National Park large enough to sustain all its living resources for all Americans in perpetuity.

The park would serve as the core area of a new Klamath-Siskiyou Biosphere Reserve to protect the biota and ecological processes of northwestern California and southwestern Oregon. Following the model of other such reserves worldwide, consumptive human uses would be excluded within the park; the rest of the reserve would be used to protect the park's resources and to study and practice sustainable management of the region's forest resources. In addition to management by individual owners within the reserve's boundaries, issues that transcend these boundaries would be dealt with cooperatively by the National Park Service, the Forest Service, the BLM, the states, and private landowners.

The Klamath-Siskiyou region is only one of the Westside areas where too little ancient forest is protected. Conservation biologists have shown unmistakably that preserved areas, even ones as large as Mt. Rainier National Park, lose species when they become isolated islands surrounded by development. Furthermore, the preponderance of land within Westside national parks and wilderness areas is not ancient forest. Aggressive efforts to enlarge existing protected areas and to preserve other areas are essential to avoid further losses. Too little ancient forest has been protected.

In keeping with the new priority, the plans must emphasize how preservation and management activities will contribute to maintaining and restoring biological diversity. Plans need to be coordinated among all twelve national forests and the five BLM districts to provide a comprehensive understanding of what we are doing throughout the Westside region.

MINIMIZING FRAGMENTATION

As pervasive as forest destruction is, fragmentation has become an even greater threat to biological diversity; forests are being fragmented even faster than they are being destroyed. Fragmentation must be addressed in all decisions concerning preservation and management.

To preserve our options, the agencies need to end all practices that fragment the larger remaining blocks of forest, those which are more likely than small patches to remain viable during coming periods of atmospheric change. Minimum fragmentation techniques, such as those already in use in the Blue River Ranger District of Willamette National Forest, need to be examined closely, because they appear to be less damaging to biological diversity and because they decrease road mileage, thereby cutting costs of road building and maintenance. But clustering that produces large bare areas also has adverse consequences. Although maximum dispersion of cuts is the worst thing happening to Westside forests, the land management agencies need to mitigate inevitable problems that will arise from clustering cuts as well.

The agencies should enlist the aid of professional societies, including the Society of American Foresters, the Society for Conservation Biology, the Ecological Society of America, the Wildlife Society and the American Fisheries Society, to develop ways to minimize and reverse the damaging effects of fragmentation in Westside forest ecosystems. The least costly way to modify forest practices to achieve major benefits for biological diversity is to minimize fragmentation.

MINIMIZING SIMPLIFICATION

Current intensive timber management practices simplify not only lands that are cut but also adjacent lands and waters that are not; the influence of logging extends beyond—downslope and downstream of the cutover area. After preserving as much forestland as necessary and minimizing fragmentation on the rest, the next important way to maintain biological diversity is to minimize simplification.

This means, among other things, that thinning operations need to be nonselective in terms of species; they should not discriminate against species that are not valued for timber. Large snags and logs need to be present on the site throughout succession. Natural regeneration must be encouraged, with artificial planting of local stock only in exceptional cases. Succession must be allowed to provide species and structural diversity. Existing even-age monocultures should be manipulated to resemble natural stands more closely, to increase structural complexity, and to provide for managed forests having trees of different species and ages. Natural agents of disturbance—fire, insects, and pathogens—should be managed to mimic their natural regimes in the forest ecosystem.

Foresters have already made considerable strides in snag management on an experimental basis, although implementation has lagged. The Park Service has a basically sound policy on naturally caused fires, although some modifications are warranted. These are steps in the right direction and need to be broadened to encompass other ways of minimizing simplification.

THE KNOWLEDGE BASE FOR THE NEW FORESTRY

Foresters need to become at least as knowledgeable about forest ecology and conservation biology as they are about silviculture and forest economics. Until now, forestry on public lands in the Northwest has been all but synonymous with timber management, a dominance that has distorted hiring and promotion practices, training, research, and, consequently, all forestry activities. Seeing forests as ecosystems with many values, not just as fiber factories, requires that forest managers understand how these ecosystems work.

To practice the new forestry, forestland management agencies need to recruit managers at all levels with substantial training or experience in ecology and conservation biology. A forester needs to be as familiar with population biology as with silviculture and know how to maintain nutrient cycles as well as how to lay out a timber sale. This can be accomplished in two ways: by hiring managers who come from ecology or conservation biology programs or from forestry schools that have strong programs in the new forestry, where ecology and conservation biology are fundamental components. This will encourage the revamping of all forestry school curricula to include these components.

In addition to hiring people trained in sustainable forestry, men and women who practice the new forestry must be given opportunities for advancement that are now, in reality, reserved for timber managers. Public forestland management agencies need to revise the official and unstated criteria for assignment and advancement to ensure that people who understand how to maintain the biological diversity of forestlands are managing the forests—from the district level to regional headquarters. And public land agencies need to provide managers with traditional forestry backgrounds with attractive opportunities for training in the new forestry. Since the training of agency personnel depends substantially on forestry schools, the land management agencies, forestry schools, and professional societies—including the Ecological Society of America, the Society for Conservation Biology, and the Society of American Foresters—should work jointly to develop minimum training standards for forest managers in public agencies and to revise guidelines for forestry curricula to reflect the needs of the new forestry.

In a changing world, where many essentials are unknown, research is fundamental to sound forestland management. The Forest Service has both the potential and the need to be the nation's leader in research on conservation biology and landscape ecology; it can also make major contributions in other areas of ecology, hydrology, geomorphology, and soil science. The BLM's efforts in these areas have been much more modest, although its need for more research is comparable. The National Park Service also needs a strong research program. For all these agencies, research merits a far higher priority in staffing and budgeting.

More research, however essential, is not sufficient. The Forest Service already has many fine researchers in California, Oregon, and Washington, but their findings have little effect on current management practices because researchers are "out of the loop." On-site and higher-level decisions are made by timber managers, often with little knowledge of current research findings on forest processes. The agencies need to ensure that the results of invigorated scientific research programs play a greater role in forest management at all levels.

GLOBAL CHANGE SYMPOSIUM

Westside forest ecosystems are sensitive to climate, and they in turn affect climate by storing enormous amounts of carbon. Thus, their fate is intimately tied to global climatic change. Despite the overwhelming scientific opinion that we will soon experience major climatic change, agency activities, policies, and planning are not yet responding. This is dangerously shortsighted because Westside forests and climate are so important.

The Forest Service, the BLM, the Park Service, the Fish and Wildlife Service, the Department of Energy, the National Aeronautics and Space Administration, the National Oceanic and Atmospheric Administration, and the Environmental Protection Agency should convene a symposium on global change and Westside forests to address two basic issues—the effects of global change on the forests and how various practices affect their carbon budgets—and two consequent policy questions—how to mitigate these effects and how best to use Westside forests as carbon sinks, as a part of the U.S. strategy of combating global warming. The invitees should include not only agency representatives but also experts on these issues in academia, the timber industry, and the environmental community.

ANCIENT FOREST COMMISSION

Until now, America's forests have been managed as sources of timber rather than as ecosystems that provide a broad range of indispensable products and services. As a result, the ancient forests are all but gone everywhere else in the contiguous forty-eight states and are rapidly disappearing in the Westside. Their destruction is strongly reminiscent of the destruction of whales in earlier decades; both involve overexploitation of very valuable, very large, very long-lived organisms. A fundamentally new mechanism is needed to protect America's remaining ancient forests, one that transcends existing administrative boundaries. The Marine Mammal Protection Act, which created the Marine Mammal Commission, serves as a useful model for a mechanism that can address the destruction of ancient forests.

What we need is federal legislation to create an Ancient Forest Commission. Under this legislation, the commission would be composed of three members appointed by the president, with the advice and consent of Congress. Commissioners would be selected from a list of individuals knowledgeable in forest ecology, landscape ecology, and conservation biology, people who are not in a position to profit from the logging of ancient forests. Such a list would be submitted to the president by the Council on Environmental Quality, in agreement with the Smithsonian Institution, the National Science Foundation, and the National Academy of Sciences. As with the Marine Mammal Commission, the Ancient Forest Commission would establish a Committee of Scientific Advisers with expertise in forest ecology, landscape ecology, and conservation biology. Congress would allocate funding for the commission members, the Committee of Scientific Advisers, an executive director, and sufficient support staff.

The Ancient Forest Commission would:

1. Comment officially on the adequacy of management plans for the national forests and BLM lands, programmatic EISs, federal agency documents concerning timber and salvage sales, listing petitions and recovery plans for endangered and threatened species that occur in ancient forests, the Westside forest inventory, and other planning documents relevant to ancient forests anywhere in the United States. Any recommendations made by the commission to federal officials would be answered officially within 120 days. Any commission recommendations that are not followed or adopted would be referred to the commission together with a detailed explanation of why those recommendations were not followed or adopted.

2. Help the forestland management agencies to identify priorities for protecting ancient forests.

3. Conduct research and provide the latest scientific information to decision makers in the forestland management agencies to ensure that they incorporate current developments in fields such as conservation biology and landscape ecology into planning and other decision making.

4. Issue an annual report to Congress on the status of U.S. ancient forests, important research findings and needs concerning ancient forests, and Forest Service and BLM compliance with the new priority of preventing the loss of biological diversity in the ancient forests of the Westside.

5. Present expert testimony before Congress and in court concerning ancient forests.

My hope is that these seven recommendations will provide a framework for ending the destruction of ancient forests while sustaining the production of timber in a rapidly changing world. Many details still have to be worked out—a process that will take time, effort, and cooperation—if we are to succeed. It is up to Congress, management agencies, conservationists, timber companies, and interested citizens to see that the United States takes the necessary actions to save the ancient forests of the Pacific Northwest . . . while there is still time.

Glossary

Anadromous species: one whose individuals are born in freshwaters but migrate to and feed in the sea before returning to freshwaters to breed.

Ancient forest (old-growth forest): a late successional or climax stage in forest development. In the Westside, ancient or old-growth forests have a canopy of very large living conifers, shade-tolerant trees beneath the canopy, and abundant large snags and downed logs.

Angiosperms: flowering plants; the group of vascular plants that dominates most of the world's terrestrial ecosystems.

Avifauna: the birds of a given area.

Biogeochemical processes: the chemical processes that are governed by living things and their physical settings.

Biogeography: the study of the geographic distributions of living things.

Biological diversity: the diversity of living things. Three levels of biological diversity are genetic diversity, species diversity, and ecosystem diversity.

Biomass: the amount of living or once-living material. Biomass is often expressed in grams, kilograms, or tons of dry weight per unit area.

Biome: ecosystem types that cover large fractions of continents, such as tundra, temperate coniferous forest, or tropical rain forest.

Biosphere: the thin film of land, water, and air that is home to all life on Earth.

Boreal taiga forest: the coniferous forest biome dominated by some combination of white or black spruce, jack or Scotch pine, or larch that stretches across Alaska, Canada, Scandinavia, and the USSR south of the tundra biome.

Broadleaves: trees and shrubs in the flowering plant phylum Angiospermae.

Browser: in wildlife biology, an herbivorous animal that feeds on browse (the leaves and twigs of woody plants).

Canopy: the upper level of forest vegetation, which intercepts most sunlight, rain, and snow.

287

Clearcutting: a logging technique that eliminates all or virtually all the forest canopy trees.

Climax forest: the final stage in forest succession, in which species composition exhibits no marked directional change with time.

Community: an interacting assemblage of species living in the same area. For example, a western hemlock–Douglas-fir community consists of these two tree species and many other plants, animals, and microorganisms.

Congener: a member of the same genus (qv). Silver fir (*Abies amabilis*) and noble fir (*Abies procera*) are congeneric, while Douglas-fir (*Pseudotsuga menziesii*) is not their congener.

Conifer: a tree or shrub in the phylum Gymnospermae whose seeds are borne in woody cones. There are about 500–600 species of living conifers.

Conspecific: a member of the same species.

Deciduous trees and shrubs: those whose leaves last a year or less; they drop and replace their leaves over periods sufficiently distinct that they are leafless for some portion of each year. Examples in the Pacific Northwest include angiosperms such as red alder and bigleaf maple and one gymnosperm, western larch.

Deforestation: long-term or permanent destruction of the forest.

Detritus: decomposing, formerly living material and the organisms that break it down.

Disturbance: in forest ecology, an event that eliminates some of the existing vegetation, thereby creating an opportunity for other plants. Some kinds of disturbances include volcanic eruptions, fires, landslides, and outbreaks of defoliating insects.

Duff: an organic layer atop the mineral soil consisting of fallen leaves and wood in all stages of decay and decomposer fungi, bacteria, and animals. Also called forest floor.

Ecosystem: a community of organisms and its physical setting. An ecosystem, whether a fallen log or an entire watershed, includes resident organisms, nonliving components such as soil nutrients, inputs such as rainfall, and outputs such as organisms that disperse to other ecosystems.

Ecosystem services: free services, such as providing clean air and clean water, performed by ecosystems.

Endemic: occurring within an area and nowhere else in the wild. For example, Port Orford cedar is endemic to a small area of southwest Oregon and northwest California.

Epiphyte: a plant or fungus that grows on other plants, but not as a parasite. Common Pacific Northwest epiphytes in Sitka spruce and Douglas-fir forests include green algae, liverworts, mosses, clubmosses, ferns, angiosperms, microscopic fungi, and lichens.

Evergreens: trees and shrubs whose leaves last more than one year and which drop and replace their leaves gradually rather than in sudden pulses. Examples in the Northwest include gymnosperms such as Douglas-fir and Pacific yew and angiosperms such as tanoak and madrone.

Extinction: the death of species and populations.

Family: in zoology, botany, and mycology, the taxon or group comprised of closely related genera; for example, mink and weasels (genus *Mustela*), marten and fishers (genus *Martes*), and river otters (genus *Lutra*) are members of the family Mustelidae. In tree genetics, the term "family" refers to a line of genetically similar individuals with common ancestors.

Forest management: subjecting forests or tree plantations to human manipulations, such as planting, thinning, or fire suppression. Although forest management can have many possible objectives, such as storing carbon, diminishing the risk of flooding, or increasing populations of rare species, the objective of almost all current forest management is to provide timber for logging.

Genetic diversity: the lowest level of biological diversity; the amount of genetic variation among individuals in a population and among populations of a species.

Genus: a biological classification or taxon with one or a group of closely related species. Some Westside genera include *Strix* (spotted and barred owls), *Oncorhynchus* (the five northwestern species of Pacific salmon), *Homo* (humans), *Tsuga* (western and mountain hemlocks), and *Cantharellus* (chanterelle mushrooms).

Grazer: a herbivorous animal; in wildlife biology, a herbivore feeds on grasses.

Harvest: the gathering of a cultivated crop. The term is used inappropriately by timber-oriented foresters as a euphemism for killing trees in virgin forests; it is used appropriately in reference to tree plantations, which are, in fact, cultivated crops.

Heartwood: the inner wood, found only in older trees. Heartwood supports the tree but no longer plays a role in conducting water. Because trees deposit certain metabolic by-products in heartwood, it is usually more rot-resistant than sapwood.

Herpetofauna: the amphibians and reptiles of a given area.

Hyphae: the threadlike filaments that form the mycelium of a fungus.

Hypogeous: fruiting below ground.

Keystone species: a species that influences other members of its community far out of proportion to its abundance.

Mesic: pertaining to an ecosystem of medium moisture.

Mineral soil: the layers of soil below the duff, consisting largely of mineral particles with some organic material.

Mutualism: a kind of symbiosis or intimate association between two or more organisms in which the participants benefit from their differences. Examples include red alders and the nitrogen-fixing bacteria in their root nodules.

Mycorrhizae: a mutualism between plant roots and certain kinds of fungi. The plants exude carbon compounds to the fungi, and the fungi provide the plants with soil nutrients, such as phosphorus.

National forest: lands owned by the American people and managed by the USDA Forest Service. There are currently 156 national forests, totaling some 191 million acres of land. Twelve of them, totaling 15.355 million acres, lie west of the crest of the Cascades in Washington, Oregon, and northwestern California.

Natural forest: forest whose structure, composition, and processes have not been substantially affected by human activities. There are probably no forests anywhere that are completely unaffected by burning, logging, agriculture, or atmospheric changes, but the Pacific Northwest still hosts significant amounts of relatively unaltered natural forest.

Nitrogen fixation: the process by which some bacteria and cyanobacteria (blue-green algae) that live freely or mutualistically with plants or fungi convert atmospheric nitrogen into nitrogenous compounds that can be used by plants.

Nival zone: the zone of perpetual snow, typically higher in elevation than, or poleward of, tundra.

Old-growth forest: ancient forest.

Parr: a postlarval salmon before it enters the sea.

Pathogen: an organism that causes disease in other organisms.

Photosynthates: carbon-containing compounds, such as sugars and amino acids, that plants, using light energy, manufacture from carbon dioxide and water in the process of photosynthesis.

Pioneer community: the first seral community that colonizes a site.

Population: all the interbreeding individuals of a species in a given area.

Productivity: the addition of biomass in a given area per unit time. Whereas ecologists are usually concerned with productivity of all the organisms in an ecosystem, some foresters equate productivity with addition of wood.

Recruitment: the addition of new individuals to the population.

Resilience: one kind of ecological stability, namely, the ability of the system to recover from disturbance.

Resistance: one kind of ecological stability, namely, the ability of the system to resist or tolerate disturbance.

Riparian zone: the zone surrounding a stream or river.

Rotation: the interval between cuts that eliminate the canopy. On Westside private lands, Douglas-fir stands are cut for timber as early as forty years; in national forests, 80–110 years is more common, and some longer rotations are planned.

Sapwood: the outer wood of a tree; it actively conducts water and dissolved minerals from the roots to the branches.

Seeps: small springs.

Seral community: a pioneer or any later successional community before the development of a climax community.

Shade-tolerant trees: those that can grow, survive, and (in some cases) reproduce in the shade of their own or other tree species. Examples include western hemlock, Pacific yew, and tanoak. Many shade-tolerant trees nonetheless grow, survive, and reproduce better in less shady situations.

Snag: a standing dead tree. Safety officials consider them hazardous and call them "danger trees," but many scientists and managers are now calling them "wildlife trees" in recognition of their crucial importance to wildlife and the regrowth of forests. Hard snags are newer snags whose wood is basically sound and difficult to penetrate; soft snags are easily penetrable by animals.

Species: a group of organisms that can reproduce successfully with one another but not with other organisms. Some species consist of only a single population; others have many more or less discrete populations.

Stumpage: standing timber; its value or the right to cut it.

Subnivean: under snow.

Succession: change in the species composition and structure of a community with time. Successional changes are driven by the differing abilities of organisms to colonize sites when the opportunity arises, by the environmental changes that organisms cause, and by the differing effects of environmental fluctuations, such as droughts, fires, or disease outbreaks.

Synergism: an interaction between factors whose effects exceed their additive effects. For example, increasing temperature and decreasing humidity have a synergistic effect on the risk of fire.

Transpiration: the vaporization of water from plant tissues through pores called stomata.

Trophic: pertaining to feeding, as in trophic interactions.

Vegetation management: manipulating plant communities to achieve desired objectives. In the Westside, this usually means using herbicides to kill plants that compete with Douglas-firs, with the intent of diminishing competition for light, water, and nutrients.

Virgin forest: forest that has not been logged or burned by humans.

Suggested Readings

CHAPTER 2

Arno, Stephen F., and Ramona P. Hammerly (1977). *Northwest Trees.* The Mountaineers, Seattle, WA.

Bernstein, Art (1986). *Trees of Southern Oregon.* New Leaf Books, Grants Pass, OR.

Brokaw, Howard P. (1978). *Wildlife and America.* Council on Environmental Quality, Washington, DC.

Brubaker, Linda B. (in press). "Climatic Change and the Origin of Douglas-Fir/ Western Hemlock Forests in the Puget Sound Lowlands." In *Proceedings from Old-Growth Douglas-Fir Forests: Wildlife Communities and Habitat Relationships, a Symposium March 29, 30, 31, 1989, Portland, Oregon.*

Cox, Thomas R., Robert S. Maxwell, Phillip Drennon Thomas, and Joseph J. Malone (1985). *This Well-Wooded Land: Americans and Their Forests from Colonial Times to the Present.* University of Nebraska Press, Lincoln, NE.

Daubenmire, Rexford (1978). *Plant Geography.* Academic Press, New York, NY.

Elias, Thomas S. (1980). *The Complete Trees of North America: Field Guide and Natural History.* Outdoor Life/Nature Books, New York, NY.

Fowells, H. A. (1965). *Silvics of Forest Trees of the United States.* USDA Forest Service Agriculture Handbook no. 271. Washington, DC.

Franklin, Jerry F., Kermit Cromack, Jr., William Denison, Arthur McKee, Chris Maser, James Sedell, Fred Swanson, and Glenn Juday (1981). *Ecological Characteristics of Old-Growth Douglas-Fir Forests.* USDA Forest Service, Pacific Northwest Forest and Range Experiment Station General Technical Report PNW-118, Portland, OR.

Franklin, Jerry F., and C. T. Dyrness (1973). *Natural Vegetation of Oregon and Washington.* USDA Forest Service, Pacific Northwest Forest and Range Experiment Station General Technical Report PNW-8, Portland, OR.

Geppert, Rollin R., Charles W. Lorenz, and Arthur G. Larson (1984). *Cumulative Effects of Forest Practices on the Environment: A State of the Knowledge.* Ecosystems, Inc., Olympia, WA.

Gunther, Erna (1945). *Ethnobotany of Western Washington: The Knowledge and Use of Indigenous Plants by Native Americans.* University of Washington Press, Seattle, WA.

Kelly, David, and Gary Braasch (1988). *Secrets of the Old Growth Forest.* Gibbs Smith, Salt Lake City, UT.

Kimmerling, A. Jon, and Philip L. Jackson, eds. (1985). *Atlas of the Pacific Northwest,* 7th ed. Oregon State University Press, Corvallis, OR.

Olson, Jeffrey T. (1988). *National Forests: Policies for the Future.* Vol. 4: *Pacific Northwest Lumber and Wood Products: An Industry in Transition.* The Wilderness Society, Washington, DC.

Omernik, James M., and Alisa L. Gallant (1986). *Ecoregions of the Pacific Northwest*. EPA/600/3–86/033, U.S. Environmental Protection Agency, Environmental Research Laboratory, Corvallis, OR.

Pierre, Joseph H. (1979). *When Timber Stood Tall*. Superior Publishing Co., Seattle, WA.

Ruby, Robert H., and John A. Brown (1986). *A Guide to the Indian Tribes of the Pacific Northwest*. University of Oklahoma Press, Norman, OK.

Rushforth, Keith D. (1987). *Conifers*. Christopher Helm, London.

Shumway, Stuart E. (1981). "Climate." Pp. 87–92, in Paul E. Heilman, Harry W. Anderson, and David M. Baumgartner, eds. In *Forest Soils of the Douglas-Fir Region*. Washington State University, Pullman, WA.

Underhill, Ruth (1944). *Indians of the Pacific Northwest*. U.S. Office of Indian Affairs, Washington, DC.

Van Gelderen, D. M. (1986). *Conifers*. Timber Press, Portland, OR.

Visher, Stephen S. (1954). *Climatic Atlas of the United States*. Harvard University Press, Cambridge, MA.

Waring, R. H., and J. F. Franklin (1979). "Evergreen Coniferous Forests of the Pacific Northwest." *Science* 204:1380–86.

Whitney, Stephen R. (1983). *A Field Guide to the Cascades and Olympics*. The Mountaineers, Seattle, WA.

Williams, Richard L. (1976). *The Loggers*. Time-Life Books, New York, NY.

CHAPTER 3

Anderson, H. Michael, and Deanne Kloepfer (1988). *End of the Ancient Forests: Special Report on the National Forest Plans in the Pacific Northwest*. The Wilderness Society, Washington, DC.

Bartels, Ronald, John D. Dell, Richard L. Knight, and Gail Schaefer (1985). "Dead and Down Woody Material." Pp. 171–86, in E. Reade Brown, ed. *Management of Wildlife and Fish Habitats in Forests of Western Oregon and Washington, Part 1—Chapter Narratives*. USDA Forest Service, Pacific Northwest Region, Portland, OR.

Bisson, Peter A., Robert E. Bilby, Mason D. Bryant, C. Andrew Dolloff, Glenn B. Grette, Robert A. House, Michael L. Murphy, K. Victor Koski, and James R. Sedell (1987). "Large Woody Debris in Forested Streams in the Pacific Northwest: Past, Present, and Future." Pp. 143–90, in Ernest O. Salo and Terrence W. Cundy, eds. *Streamside Management: Forestry and Fishery Interactions*. Contribution no. 57. University of Washington Institute of Forest Resources, Seattle, WA.

Botkin, Daniel B. (1980). "A Grandfather Clock Down the Staircase: Stability and Disturbance in Natural Ecosystems." Pp. 1–10, in Richard H. Waring, ed. *Forests: Fresh Perspectives from Ecosystem Analysis*. Oregon State University Press, Corvallis, OR.

Carroll, George C. (1980). "Forest Canopies: Complex and Independent Subsystems." Pp. 87–107, in Richard H. Waring, ed. *Forests: Fresh Perspectives from Ecosystem Analysis*. Oregon State University Press, Corvallis, OR.

Cline, S. P., A. B. Berg, and H. M. Wight (1980). "Snag Characteristics and Dynamics in Douglas-Fir Forests, Western Oregon." *Journal of Wildlife Management* 44(4):773–86.

Conard, Susan G., Annabelle E. Jaramillo, Kermit Cromack, Jr., and Sharon

Rose, eds. (1985). *The Role of the Genus* Ceanothus *in Western Forest Ecosystems.* USDA Forest Service, Pacific Northwest Forest and Range Experimental Station General Technical Report PNW-182, Portland, OR.

Franklin, Jerry F., Kermit Cromack, Jr., William Denison, Arthur McKee, Chris Maser, James Sedell, Fred Swanson, and Glenn Juday (1981). *Ecological Characteristics of Old-Growth Douglas-Fir Forests.* USDA Forest Service, Pacific Northwest Forest and Range Experiment Station General Technical Report PNW-118, Portland, OR.

Franklin, Jerry F., and C. T. Dyrness (1973). *Natural Vegetation of Oregon and Washington.* USDA Forest Service, Pacific Northwest Forest and Range Experiment Station General Technical Report PNW-8, Portland, OR.

Franklin, Jerry F., and Miles A. Hemstrom (1981). "Aspects of Succession of the Coniferous Forests of the Pacific Northwest." Pp. 212–29, in Darrell C. West, Herman H. Shugart and Daniel B. Botkin, eds. *Forest Succession: Concepts and Application.* Springer-Verlag, New York, NY.

Franklin, Jerry F., H. H. Shugart, and Mark E. Harmon (1987). "Tree Death as an Ecological Process." *BioScience* 37(8):550–56.

Graham, Robin Lee Lambert (1981). "Biomass Dynamics of Dead Douglas-Fir and Western Hemlock Boles in Mid-Elevation Forests of the Cascade Range." Ph.D. dissertation, Oregon State University, Corvallis, OR.

Harmon, Mark E. (1987). "The Influence of Litter and Humus Accumulations and Canopy Openness on *Picea sitchensis* (Bong.) Carr. and *Tsuga heterophylla* (Raf.) Sarg. Seedlings Growing on Logs." *Canadian Journal of Forest Research* 17:1475–79.

Harmon, M. E., J. F. Franklin, F. J. Swanson, P. Sollins, S. V. Gregory, J. D. Lattin, N. H. Anderson, S. P. Cline, N. G. Aumen, J. R. Sedell, G. W. Lienkaemper, K. Cromack, Jr., and K. W. Cummins (1986). "Ecology of Coarse Woody Debris in Temperate Ecosystems." *Advances in Ecological Research* 15:133–302.

Harris, Larry D. (1984). *The Fragmented Forest: Island Biogeography Theory and the Preservation of Biotic Diversity.* University of Chicago Press, Chicago, IL.

Kimmins, J. P. (1987). *Forest Ecology.* Macmillan Publishing Co., New York, NY.

Maser, Chris, Robert F. Tarrant, James M. Trappe, and Jerry F. Franklin, technical eds. (1988). *From the Forest to the Sea: A Story of Fallen Trees.* USDA Forest Service, Pacific Northwest Research Station General Technical Report PNW-GTR-229, Portland, OR.

Maser, Chris, James M. Trappe, Steven P. Cline, Kermit Cromack, Jr., Helmut Blaschke, James R. Sedell, and Frederick J. Swanson (1984). *The Seen and Unseen World of the Fallen Tree.* USDA Forest Service, Pacific Northwest Forest and Range Experimental Station General Technical Report PNW-164, Portland, OR.

Maser, Chris, James M. Trappe, and Ronald A. Nussbaum (1978). "Fungal-Small Mammal Interrelationships with Emphasis on Oregon Coniferous Forests." *Ecology* 59(4)799–809.

Means, Joseph E., ed. (1982). *Forest Succession and Stand Development Research in the Northwest.* Forest Research Laboratory, Oregon State University, Corvallis, OR.

Nietro, William A., Virgil W. Binkley, Steven P. Cline, R. William Mannan, Bruce G. Marcot, Douglas Taylor, and Frank F. Taylor (1985). "Snags (Wildlife Trees)." Pp. 129–69, and Appendices 18 and 19, in E. Reade Brown, ed. *Man-*

agement of Wildlife and Fish Habitats in Forests of Western Oregon and Washington, Part 1—Chapter Narratives and Part 2—Appendices. USDA Forest Service, Pacific Northwest Region, Portland, OR.

Old-Growth Definition Task Group (J. F. Franklin, F. Hall, W. Laudenslayer, C. Maser, J. Nunan, J. Poppino, C. J. Ralph, and T. Spies) (1986). *Interim Definitions for Old-Growth Douglas-Fir and Mixed-Conifer Forests in the Pacific Northwest and California.* USDA Forest Service, Pacific Northwest Research Station Research Note PNW-447, Portland, OR.

Perry, D. A., R. Molina, and M. P. Amaranthus (1987). "Mycorrhizae, Mycorrhizospheres, and Reforestation: Current Knowledge and Research Needs." *Canadian Journal of Forest Research* 17(8):929–40.

Pickett, S. T. A., and P. S. White, eds. (1985). *The Ecology of Natural Disturbance and Patch Dynamics.* Academic Press, Inc., Orlando, FL.

Pike, Lawrence H., Robert A. Rydell, and William C. Denison (1977). "A 400-Year-Old Douglas-Fir Tree and Its Epiphytes: Biomass, Surface Area, and Their Distributions." *Canadian Journal of Forestry* 7(4):680–99.

Sollins, Phillip, Steven P. Cline, Thomas Verhoeven, Donald Sachs, and Gody Spycher (1987). "Patterns of Log Decay in Old-Growth Douglas-Fir Forests." *Canadian Journal of Forest Research* 17(12):1585–95.

Sollins, P., C. C. Grier, F. M. McCorison, K. Cromack, Jr., R. Fogel, and R. L. Fredriksen (1980). "The Internal Element Cycles of an Old-Growth Douglas-Fir Ecosystem in Western Oregon." *Ecological Monographs* 50(3):261–85.

Spies, Thomas A., and Jerry F. Franklin (1988). "Old Growth and Forest Dynamics in the Douglas-Fir Region of Western Oregon and Washington." *Natural Areas Journal* 8(3):190–201.

Trappe, James M., and Chris Maser (1977). "Ectomycorrhizal Fungi: Interactions of Mushrooms and Truffles with Beasts and Trees." Pp. 165–79, in Tony Walters, ed. *Mushrooms and Man: An Interdisciplinary Approach to Mycology.* Lynn-Benton Community College, Albany, OR.

Vogt, K. A., C. C. Grier, C. E. Meier, and M. R. Keyes (1983). "Organic Matter and Nutrient Dynamics in Forest Floors of Young and Mature *Abies amabilis* Stands in Western Washington, as Affected by Fine-Root Input." *Ecological Monographs* 53:139–57.

Waring, R. H. (1987). "Characteristics of Trees Predisposed to Die." *BioScience* 37(8):569–74.

Waring, R. H., and J. F. Franklin (1979). "Evergreen Coniferous Forests of the Pacific Northwest." *Science* 204:1380–86.

CHAPTER 4

Abrams, Leroy (1923–60). *Illustrated Flora of the Pacific States, Washington, Oregon, and California.* 4 vols. Stanford University Press, Stanford, CA.

Acker, R. 1986. *The Harvest of Wild Mushrooms in Washington: An Issue Paper.* Graduate School of Public Affairs, University of Washington, Seattle, WA.

Arno, Stephen F., and Ramona P. Hammerly (1977). *Northwest Trees.* The Mountaineers, Seattle, WA.

Barker, Karlyn (1986). "The Mechanical Pterosaurus Wrecks; Thousands Watch as Crash Ends Performance of Reptile Replica." *Washington Post*, May 18, 1986, pp. A1, A17.

Beschta, Robert, Robert E. Bilby, George W. Brown, L. Blair Holtby, and Terry D. Hofstra (1987). "Stream Temperature and Aquatic Habitat: Fisheries and Forestry Interactions." Pp. 191–232, in Ernest O. Salo and Terrence W. Cundy, eds. *Streamside Management: Forestry and Fishery Interactions.* Contribution no. 57. University of Washington Institute of Forest Resources, Seattle, WA.

Bisson, Peter A., Robert E. Bilby, Mason D. Bryant, C. Andrew Dolloff, Glenn B. Grette, Robert A. House, Michael L. Murphy, K. Victor Koski, and James R. Sedell (1987). "Large Woody Debris in Forested Streams in the Pacific Northwest: Past, Present, and Future." Pp. 143–90, in Ernest O. Salo and Terrence W. Cundy, eds. *Streamside Management: Forestry and Fishery Interactions.* Contribution no. 57. University of Washington Institute of Forest Resources, Seattle, WA.

Bisson, Peter A., and James R. Sedell (1984). "Salmonid Populations in Streams in Clearcut vs. Old-Growth Forests of Western Washington." Pp. 121–29, in William R. Meehan, Theodore R. Merrell, Jr., and Thomas A. Hanley, eds. *Fish and Wildlife Relationships in Old-Growth Forests.* American Institute of Fishery Research Biologists, Morehead City, NC.

Booth, William (1987). "Combing the Earth for Cures to Cancer, AIDS." *Science* 237:969–70.

Brown, E. Reade (1985). *Management of Wildlife and Fish Habitats in Forests of Western Oregon and Washington.* USDA Forest Service, Pacific Northwest Region Publication R6-F&WL-192–1985, Portland, OR.

Bunnell, Fred L., and Greg W. Jones (1984). "Black-Tailed Deer and Old-Growth Forests—a Synthesis." Pp. 411–20, in William R. Meehan, Theodore R. Merrell, Jr., and Thomas A. Hanley, eds. *Fish and Wildlife Relationships in Old-Growth Forests.* American Institute of Fishery Research Biologists, Morehead City, NC.

Bury, R. Bruce (1988). "Habitat Relationships and Ecological Importance of Amphibians and Reptiles." Pp. 61–76, in K. J. Raedeke, ed. *Streamside Management: Riparian Wildlife and Forestry Interactions.* Contribution no. 59. Institute of Forest Resources, University of Washington, Seattle, WA.

Bury, R. Bruce, and Paul Stephen Corn (1988). "Douglas-Fir Forests in the Oregon and Washington Cascades: Relationship of the Herpetofauna to Stand Age and Moisture." Pp. 11–22, in Robert C. Szaro, Kieth R. Severson, and David R. Patton, technical coordinators. *Management of Amphibians, Reptiles, and Small Mammals in North America.* USDA Forest Service General Technical Report RM-166, Rocky Mountain Forest and Range Experiment Station, Ft. Collins, CO.

Carroll, G. C. 1988. "Fungal Endophytes in Stems and Leaves: From Latent Pathogen to Mutualistic Symbiont." *Ecology* 69(1):2–9.

Carroll, G. C., and F. E. Carroll (1978). "Studies on the Incidence of Coniferous Needles Endophytes in the Pacific Northwest." *Canadian Journal of Botany* 56:3034–43.

Cooke, W. B. (1949). "*Oxyporus nobilissimus* and the Genus *Oxyporus* in North America." *Mycologia* 41:442–55.

Corn, Paul Stephen, and R. Bruce Bury (1989). "Logging in Western Oregon: Responses of Headwater Habitats and Stream Amphibians." *Forest Ecology and Management* 29(01):1–19.

Corn, Paul Stephen, R. Bruce Bury, and Thomas A. Spies (1988). "Douglas-Fir

Forests in the Cascade Mountain of Oregon and Washington: Is the Abundance of Small Mammals Related to Stand Age and Moisture?" Pp. 340–52, in Robert C. Szaro, Kieth R. Severson, and David R. Patton, technical coordinators. *Management of Amphibians, Reptiles, and Small Mammals in North America.* USDA Forest Service, General Technical Report RM-166, Rocky Mountain Forest and Range Experiment Station, Ft. Collins, CO.

Dawson, W. R., J. D. Ligon, J. R. Murphy, J. P. Myers, D. Simberloff, and J. Verner (1986). *Report of the Advisory Panel on the Spotted Owl.* Audubon Conservation Report no. 7. National Audubon Society, New York, NY.

Denison, W. C. (1973). "Life in Tall Trees." *Scientific American* 228:74–80.

Denison, W. C. (1979). "*Lobaria oregana*, a Nitrogen-Fixing Lichen in Old-Growth Douglas-Fir Forests." Pp. 266–75, in J. C. Gordon, C. T. Wheeler, and D. A. Perry, eds. *Symbiotic Nitrogen Fixation in the Management of Temperate Forests.* Forestry Research Laboratory, Oregon State University, Corvallis, OR.

Devall, Bill, and George Sessions (1985). *Deep Ecology.* Gibbs M. Smith, Inc., Peregrine Smith Books, Salt Lake City, UT.

Dixon, Kenneth R., and Thomas C. Juelson (1987). "The Political Economy of the Spotted Owl." *Ecology* 68(7):772–76.

Duellman, William E., and Linda Trueb (1986). *Biology of Amphibians.* McGraw-Hill Book Co., New York, NY.

Ehrenfeld, David (1978). *The Arrogance of Humanism.* Oxford University Press, New York, NY.

Ehrlich, Paul, and Anne Ehrlich (1981). *Extinction: The Causes and Consequences of the Disappearance of Species.* Random House, New York, NY.

Erwin, Terry L. (1984). "Tropical Forests: Their Richness in Coleoptera and Other Arthropod Species." *Coleoptera Bulletin* 36(1):74–75.

Everest, Fred H., Neil B. Armantrout, Steven M. Keller, William D. Parante, James R. Sedell, Thomas E. Nickelson, James M. Johnston, and Gordon N. Haugen (1985). "Salmonids." Pp. 199–230, in E. Reade Brown, ed. *Management of Wildlife and Fish Habitats in Forests of Western Oregon and Washington, Part 1—Chapter Narratives.* USDA Forest Service, Pacific Northwest Region, R6-F&WL-192–1985, Portland, OR.

Fogel, R. 1980. "Mycorrhizae and Nutrient Cycling in Natural Forest Ecosystems." *New Phytologist* 86:199–212.

Fogel, R., and J. M. Trappe (1978). "Fungus Consumption (Mycophagy) by Small Animals." *Northwest Science* 52:1–31.

Forsman, E. D., and E. C. Meslow (1986). "The Spotted Owl." Pp. 743–61, in A. S. Eno, R. L. DiSilvestro, and W. J. Chandler, eds. *Audubon Wildlife Report 1986.* National Audubon Society, New York, NY.

Forsman, E. D., E. C. Meslow, and H. M. Wight (1980). "Distribution and Biology of the Spotted Owl in Oregon." *Wildlife Monographs* 87:1–64.

Fredriksen, Richard L., and R. D. Harr (1981). "Soil, Vegetation and Watershed Management." Pp. 231–60, in Paul E. Heilman, Harry W. Anderson, and David M. Baumgartner, eds., *Forest Soils of the Douglas-Fir Region.* Washington State University, Pullman, WA.

General Accounting Office (1989). *Endangered Species: Spotted Owl Petition Evaluation Beset by Problems.* GAO/RCED-89-79. Washington, DC.

Gould, G. I., Jr. (1977). "Distribution of the Spotted Owl in California." *Western Birds* 8(4):131–46.

Suggested Readings 299

Hadley, Jane (1987). "Lowly NW Tree Sought for Cancer Fight." *Seattle Post-Intelligencer*, April 20, 1987, pp. D1, D3.

Hager, Donald C. (1960). "The Interrelationships of Logging, Birds, and Timber Regeneration in the Douglas-Fir Region of Northwestern California." *Ecology* 41(1):116–25.

Hansen, E. M. (1979). "Survival of *Phellinus wierii* in Douglas-Fir Stumps After Logging." *Canadian Journal of Forest Research* 9:484–88.

Harley, J. L., and S. E. Smith (1983). *Mycorrhizal Symbiosis*. Academic Press, New York, NY.

Harmon, M. E., J. F. Franklin, F. J. Swanson, P. Sollins, S. V. Gregory, J. D. Lattin, N. H. Anderson, S. P. Cline, N. G. Aumen, J. R. Sedell, G. W. Lienkaemper, K. Cromack, Jr., and K. W. Cummins (1986). "Ecology of Coarse Woody Debris in Temperate Ecosystems." *Advances in Ecological Research* 15:133–302.

Harper, James A. (and colleagues) (1987). *Ecology and Management of Roosevelt Elk in Oregon*. Oregon Department of Fish and Wildlife.

Harris, Larry D., and Chris Maser (1984). "Animal Community Characteristics." Pp. 44–68, in Larry D. Harris, *The Fragmented Forest: Island Biogeography Theory and the Preservation of Biotic Diversity*. University of Chicago Press, Chicago, IL.

Hartzell, Hal, and Jerry Rust (1983). *Yew.* Hal Hartzell and Jerry Rust, Eugene, OR.

Hitchcock, C. L., A. Cronquist, M. Ownbey, and J. W. Thompson (1955–69). *Vascular Plants of the Pacific Northwest*. 5 parts. University of Washington Press, Seattle, WA.

Ingles, Lloyd G. (1965). *Mammals of the Pacific States*. Stanford University Press, Stanford, CA.

Jones, Greg W., and Fred R. Bunnell (1984). "Response of Black-Tailed Deer to Winters of Different Severity on Northern Vancouver Island." Pp. 385–90, in William R. Meehan, Theodore R. Merrell, Jr., and Thomas A. Hanley, eds. *Fish and Wildlife Relationships in Old-Growth Forests*. American Institute of Fishery Research Biologists, Morehead City, NC.

Manuwal, David A., and Mark H. Huff (1987). "Spring and Winter Bird Populations in a Douglas-Fir Forest Sere." *Journal of Wildlife Management* 51(3):586–95.

Marcot, Bruce G., ed. (1979). *California Wildlife Habitat Relationships Program: North Coast/Cascades Zone*. Vol. 1–4. USDA Forest Service, Six Rivers National Forest, Eureka, CA.

Mariani, Jina Marie (1987). "Brown Creeper (*Certhia americana*) Abundance Patterns and Habitat Use in the Southern Washington Cascades." M.A. thesis, University of Washington, Seattle, WA.

Marshall, David B. (1988). *Status of the Marbled Murrelet in North America with Special Emphasis on Populations in Washington, Oregon, and California*. Audubon Society of Portland, Portland, OR.

Maser, Chris, Bruce Mate, Jerry F. Franklin, and C. T. Dyrness (1981). *Natural History of Oregon Coast Mammals*. FS&BLM General Technical Report PNW-133, Portland, OR.

Maser, Chris, James M. Trappe, and Ronald A. Nussbaum (1978). "Fungal-Small Mammal Interrelationships with Emphasis on Oregon Coniferous Forests." *Ecology* 59(4):799–809.

Murphy, Michael L., and James D. Hall (1981). "Varied Effects of Clearcut Logging on Predators and Their Habitat in Small Streams of the Cascade Mountains, Oregon." *Canadian Journal of Fisheries and Aquatic Science* 38: 137–45.

Myers, Norman (1979). *The Sinking Ark: A New Look at the Problem of Disappearing Species.* Pergamon Press, New York, NY.

Myren, Richard T., and Robert J. Ellis (1984). "Evapotranspiration in Forest Succession and Long-Term Effects on Fishery Resources: A Consideration for Management of Old-Growth Forests." Pp. 183–86, in William R. Meehan, Theodore R. Merell, Jr., and Thomas A. Hanley, eds. *Fish and Wildlife Relationships in Old-Growth Forests.* American Institute of Fishery Research Biologists, Morehead City, NC.

Norse, Elliott A., and Roger E. McManus (1980). "Ecology and Living Resources: Biological Diversity." Pp. 31–80, in *Environmental Quality 1980: The Eleventh Annual Report of the Council on Environmental Quality,* Washington, DC.

Norse, Elliott A., Kenneth L. Rosenbaum, David S. Wilcove, Bruce A. Wilcox, William H. Romme, David W. Johnston, and Martha L. Stout (1986). *Conserving Biological Diversity in Our National Forests.* The Wilderness Society, Washington, DC.

Norton, Bryan G., ed. (1986). *The Preservation of Species.* Princeton University Press, Princeton, NJ.

Office of Technology Assessment (1987). *Technologies to Maintain Biological Diversity.* Congress of the United States, Office of Technology Assessment, Washington, DC.

Peck, Morton E. (1961). *A Manual of the Higher Plants of Oregon.* 2d ed. Binfords and Mort, Portland, OR.

Perry, D. A., M. P. Amaranthus, J. G. Borchers, S. L. Borchers, and R. E. Brainerd (1989). "Bootstrapping in Ecosystems." *BioScience* 39(4):230–37.

Pike, L. H. (1978). "The Importance of Epiphytic Lichens in Mineral Cycling." *Bryologist* 81:247–57.

Raedeke, Kenneth J., and Richard D. Taber (1982). "Mechanisms of Population Regulation in Western Washington Forests for *Cervus* and *Odocoileus.*" *Transactions of the International Congress of Game Biologists* 14:69–80.

Raphael, Martin G. (1984). "Wildlife Populations in Relation to Stand Age and Area in Douglas-Fir Forest of Northwestern California." Pp. 259–74, in William R. Meehan, Theodore R. Merrell, Jr., and Thomas A. Hanley, eds. *Fish and Wildlife Relationships in Old-Growth Forests.* American Institute of Fishery Research Biologists, Morehead City, NC.

Raphael, Martin G. (1987a). "Wildlife-Tanoak Associations in Douglas-Fir Forests of Northwestern California." Pp. 183–89, in Timothy R. Plumb and Norman H. Pillsbury, technical coordinators. *Proceedings of the Symposium on Multiple-Use Management of California's Hardwood Resources.* USDA Forest Service, Pacific Southwest Forest and Range Experiment Station General Technical Report PSW-100, Berkeley, CA.

Raphael, Martin G. (1987b). "Use of Pacific Madrone by Cavity-Nesting Birds." Pp. 198–202, in Timothy R. Plumb and Norman Pillsbury, technical coordinators. *Proceedings of the Symposium on Multiple-Use Management of California's Hardwood Resources.* USDA Forest Service, Pacific Southwest Forest and Range Experiment Station General Technical Report PSW-100, Berkeley, CA.

Raphael, Martin G. (1988). "Long-Term Trends in Abundance of Amphibians, Reptiles, and Mammals in Douglas-Fir Forests of Northwestern California."

Pp. 23–31, in Robert C. Szaro, Kieth R. Severson, and David R. Patton, technical coordinators. *Management of Amphibians, Reptiles, and Small Mammals in North America.* USDA Forest Service, General Technical Report RM-166, Rocky Mountain Forest and Range Experiment Station, Ft. Collins, CO.

Raphael, Martin G., and Reginald H. Barrett (1984). "Diversity and Abundance of Wildlife in Late Successional Douglas-Fir Forests." Pp. 352–60, in *New Forests for a Changing World, Proceedings of the 1983 SAF National Convention,* Portland, OR.

Regelin, Wayne L. (1979). "Nutritional Interactions of Black-Tailed Deer with Their Habitat in Southeast Alaska." Pp. 60–68, in O. C. Wallmo and J. W. Schoen, eds. *Sitka Black-Tailed Deer: Proceedings of a Conference in Juneau, Alaska.* USDA Forest Service, Alaska Region, Juneau. Series no. R10-48.

Ruggiero, Leonard F., Richard S. Holthausen, Bruce G. Marcot, Keith B. Aubry, Jack Ward Thomas, and E. Charles Meslow (1988). "Ecological Dependency: The Concept and Its Implications for Research and Management." Pp. 115–26, in Richard E. McCabe, ed., *Transactions of the 53rd North American Wildlife and Natural Resources Conference.* Wildlife Management Institute, Washington, DC.

Salwasser, Hal (1986). "Conserving a Regional Spotted Owl Population." Pp. 227–47, in National Research Council Committee on the Applications of Ecological Theory to Environmental Problems. *Ecological Knowledge and Environmental Problem-Solving.* National Academy Press, Washington, DC.

Schoen, John W., Matthew D. Kirchoff, and Olof C. Wallmo (1984). "Sitka Black-Tailed Deer/Old-Growth Relationships in Southeast Alaska: Implications for Management." Pp. 315–19, in William R. Meehan, Theodore R. Merrell, Jr., and Thomas A. Hanley, eds. *Fish and Wildlife Relationshps in Old-Growth Forests.* American Institute of Fishery Research Biologists, Morehead City, NC.

Schoen, John W., and O. C. Wallmo (1979). "Timber Management and Deer in Southeast Alaska: Current Problems and Research Direction." Pp. 69–85, in O. C. Wallmo and J. W. Schoen, eds. *Sitka Black-Tailed Deer: Proceedings of a Conference in Juneau, Alaska.* USDA Forest Service, Alaska Region, Juneau. Series no. R10-48.

Schoen, John W., Olof C. Wallmo, and Matthew D. Kirchoff (1981). "Wildlife-Forest Relationships: Is a Reevaluation of Old Growth Necessary?" Pp. 531–44, in *Transactions of the 46th North American Wildlife and Natural Resources Conference, 1981.* Wildlife Management Institute, Washington, DC.

Sedell, James R., and Frederick J. Swanson (1984). "Ecological Characteristics in Old-Growth Forests in the Pacific Northwest." Pp. 9–16, in William R. Meehan, Theodore R. Merrell, Jr., and Thomas A. Hanley, eds. *Fish and Wildlife Relationships in Old-Growth Forests.* American Institute of Fishery Research Biologists, Morehead City, NC.

Siddall, Jean L., Kenton L. Chambers, and David H. Wagner (1979). *Rare, Threatened, and Endangered Vascular Plants in Oregon—An Interim Report.* Oregon Natural Area Preserves Advisory Committee, State Land Board, Salem, OR.

Simberloff, Daniel (1987). "The Spotted Owl Fracas: Mixing Academic, Applied, and Political Ecology." *Ecology* 68(4):766–72.

Spies, Thomas A. (in press). "Characteristics of Old-Growth Douglas-Fir Plant Communities in Washington and Oregon." *Proceedings from Old-Growth*

Douglas-Fir Forests: Wildlife Communities and Habitat Relationships, a Symposium March 29, 30, 31, 1989, Portland, Oregon.

Stevenson, Susan K., and James A. Rochelle (1984). "Lichen Litterfall—Its Availability and Utilization by Black-Tailed Deer." Pp. 391–96, in William R. Meehan, Theodore R. Merrell, Jr., and Thomas A. Hanley, eds. *Fish and Wildlife Relationships in Old-Growth Forests.* American Institute of Fishery Research Biologists, Morehead City, NC.

Suffness, Matthew (1985). "The Discovery and Development of Antitumor Drugs from Natural Products." Pp. 101–33, in Arnold J. Vlietinck and Roger A. Dommisse, eds. *Advances in Medicinal Plant Research.* Wissenschaftliche Verlagsgesellschaft mbH, Stuttgart.

Taber, Richard D., and Thomas A. Hanley (1979). "The Black-Tailed Deer and Forest Succession in the Pacific Northwest." Pp. 33–52, in O. C. Wallmo and J. W. Schoen, eds. *Sitka Black-Tailed Deer: Proceedings of a Conference in Juneau, Alaska.* USDA Forest Service, Alaska Region, Juneau. Series no. R10-48.

Temple, S. A. (1977). "Plant-Animal Mutualism: Coevolution with Dodo Leads to Near Extinction of Plant." *Science* 197:885–86.

Thomas, Donald W. (1988). "The Distribution of Bats in Different Ages of Douglas-Fir Forests." *Journal of Wildlife Management* 54(4):619–26.

Thomas, Jack Ward, technical ed. (1979). *Wildlife Habitats in Managed Forests, the Blue Mountains of Oregon and Washington.* USDA Forest Service, Agriculture Handbook no. 553, Portland, OR.

Thomas, Jack Ward, and Larry D. Bryant (1987). "The Elk." Pp. 494–507, in Roger L. Di Silvestro, ed. *Audubon Wildlife Report 1987.* Academic Press, Inc., Orlando, FL.

Thomas, Jack Ward, Leonard F. Ruggiero, R. William Mannan, John W. Schoen, and Richard A. Lancia (1988). "Management and Conservation of Old-Growth Forests in the United States." *Wildlife Society Bulletin* 16:252–62.

Trappe, J. M., and R. Fogel (1978). "Ecosystematic Function of Mycorrhizae." Pp. 205–14, in S. A. Marshall, ed. *The Belowground Symposium: A Synthesis of Plant-Associated Processes.* Colorado State University Science Series 26, Ft. Collins, CO.

Ure, D. C., and Chris Maser (1982). "Mycophagy in Red-Backed Voles in Oregon and Washington." *Canadian Journal of Forestry* 60:3307–15.

USDA Forest Service (1984). *Regional Guide for Pacific Northwest Region.* USDA Forest Service, Northwest Region, Portland, OR.

USDA Forest Service (1988). *Final Supplement to the Environmental Impact Statement for an Amendment to the Pacific Northwest Regional Guide. Spotted Owl Guidelines.* 2 vols. USDA Forest Service, Washington, DC.

Wallmo, Olof C., ed. (1981). *Mule and Black-Tailed Deer of North America.* University of Nebraska Press, Lincoln, NE.

Waring, R. H., and J. F. Franklin (1979). "Evergreen Coniferous Forests of the Pacific Northwest." *Science* 204:1380–86.

Wilcove, David S. (1988). *National Forests: Policies for the Future. Vol. 2: Protecting Biological Diversity.* The Wilderness Society, Washington, DC.

Wilson, Edward O., ed. (1988). *Biodiversity.* National Academy Press, Washington, DC.

Witmer, Gary William (1982). "Roosevelt Elk Habitat Use in the Oregon Coast Range." Ph.D. dissertation, Oregon State University, Corvallis, OR.

Witmer, Gary W., Mike Wisdom, Edmund P. Harshman, Robert J. Anderson, Christopher Carey, Mike P. Kuttel, Ira D. Luman, James A. Rochelle, Raymond W. Scharpf, and Doug Smithey (1985). "Deer and Elk." Pp. 231–58, in E. Reade Brown, ed. *Management of Wildlife and Fish Habitats in Forests of Western Oregon and Washington, Part 1—Chapter Narratives.* USDA Forest Service, Pacific Northwest Region R6-F&WL-192–1985, Portland, OR.

CHAPTER 5

Auclair, D. (1983). " 'Natural' Mixed Forests and 'Artificial' Monospecific Forests." Pp. 71–82, in H. A. Mooney and M. Godron, eds. *Disturbance and Ecosystems: Components of Response.* Springer-Verlag, Berlin.

Berris, Steven N., and R. Dennis Harr (1987). "Comparative Snow Accumulation and Melt During Rainfall in Forested and Clearcut Plots in the Western Cascades of Oregon." *Water Resources Research* 23(1):135–42.

Beschta, Robert, Robert E. Bilby, George W. Brown, L. Blair Holtby, and Terry D. Hofstra (1987). "Stream Temperature and Aquatic Habitat: Fisheries and Forestry Interactions." Pp. 191–232, in Ernest O. Salo and Terrence W. Cundy, eds. *Streamside Management: Forestry and Fishery Interactions.* Contribution no. 57. University of Washington Institute of Forest Resources, Seattle, WA.

California Gene Resource Program (1982). *Douglas-Fir Genetic Resources: An Assessment and Plan for California.* Prepared for the State of California under the administration of the Department of Food and Agriculture, Contract no. 9146. National Council on Gene Resources, Berkeley, CA.

DeBell, Dean S., and Jerry F. Franklin (1987). "Old-Growth Douglas-Fir and Western Hemlock: A 36-Year Record of Growth and Mortality." *Western Journal of Applied Forestry* 2(4):111–14.

Dixon, Dougal (1981). *After Man: A Zoology of the Future.* Granada, London.

Ehrlich, Paul R., and Harold A. Mooney (1983). "Extinction, Substitution, and Ecosystem Services." *BioScience* 33(4):248–54.

Fogel, R. 1980. "Mycorrhizae and Nutrient Cycling in Natural Forest Ecosystems." *New Phytologist* 86:199–212.

Fogel, R., and J. M. Trappe (1978). "Fungus Consumption (Mycophagy) by Small Animals." *Northwest Science* 52:1–31.

Franklin, Jerry F. (1987). "Scientific Use of Wilderness." Pp. 42–46, in Robert C. Lucas, compiler. *Proceedings—National Wilderness Research Conference: Issues, State-of-knowledge, Future Directions; 1985 July 23–26, Ft. Collins, CO.* USDA Forest Service, Intermountain Research Station General Technical Report INT-220.

Franklin, Jerry F., and C. T. Dyrness (1973). *Natural Vegetation of Oregon and Washington.* USDA Forest Service, Pacific Northwest Forest and Range Experiment Station General Technical Report PNW-8, Portland, OR.

Fredriksen, Richard L., and R. D. Harr (1981). "Soil, Vegetation and Watershed Management." Pp. 231–60, in Paul E. Heilman, Harry W. Anderson, and David M. Baumgartner, eds., *Forest Soils of the Douglas-Fir Region.* Washington State University, Pullman, WA.

Gammon, R. H., E. T. Sundquist, and P. J. Fraser (1985). "History of Carbon Dioxide in the Atmosphere." Pp. 25–62, in J. R. Trabalka, ed. *Atmospheric Carbon Dioxide and the Global Carbon Cycle.* DOE/ER-0239, U.S. Department of Energy, Washington, DC.

Gray, Donald H., and Andrew T. Leiser (1982). *Biotechnical Slope Protection and Erosion Control.* Van Nostrand Reinhold Co., New York, NY.

Greene, Sarah (1988). "Research Natural Areas and Protecting Old-Growth Forests on Federal Lands in Western Oregon and Washington." *Natural Areas Journal* 8(1):25–30.

Harley, J. L., and S. E. Smith (1983). *Mycorrhizal Symbiosis.* Academic Press, New York, NY.

Harmon, M. E., J. F. Franklin, F. J. Swanson, P. Sollins, S. V. Gregory, J. D. Lattin, N. H. Anderson, S. P. Cline, N. G. Aumen, J. R. Sedell, G. W. Lienkaemper, K. Cromack, Jr., and K. W. Cummins (1986). "Ecology of Coarse Woody Debris in Temperate Ecosystems." *Advances in Ecological Research* 15:133–302.

Harr, R. Dennis (1980). *Streamflow After Patch Logging in a Small Drainage Within the Bull Run Municipal Watershed, Oregon.* USDA Forest Service, Research Paper PNW-268, Portland, OR.

Harr, R. D. (1981). "Some Characteristics and Consequences of Snowmelt During Rainfall in Western Oregon." *Journal of Hydrology* 53:277–304.

Harr, R. Dennis (1982). "Fog Drip in the Bull Run Municipal Watershed, Oregon." *Water Resources Bulletin* 18(5):785–89.

Harr, R. Dennis (1983). "Potential for Augmenting Water Yield Through Forest Practices in Western Washington and Western Oregon." *Water Resources Bulletin* 19(3):383–93.

Harr, R. Dennis (1986). "Effects of Clearcutting on Rain-on-Snow Runoff in Western Oregon: A New Look at Old Studies." *Water Resources Research* 27(7):1095–1100.

Heilman, Paul E., Harry W. Anderson, and David M. Baumgartner (1981). *Forest Soils of the Douglas-Fir Region.* Washington State University, Pullman, WA.

Hermann, Richard K., and Denis P. Lavander (1968). "Early Growth of Douglas-Fir from Various Altitudes and Aspects in Southern Oregon." *Silvae Genetica* 17:143–51.

Houghton, R. A., W. H. Schlesinger, S. Brown, and J. F. Richards (1985). "Carbon Dioxide Exchange Between the Atmosphere and Terrestrial Ecosystems." Pp. 113–40, in J. R. Trabalka, ed. *Atmospheric Carbon Dioxide and the Global Carbon Cycle.* DOE/ER-0239, U.S. Department of Energy, Washington, DC.

Kulp, J. Laurence (1986). "Projections for the Year 2020 in the Douglas-Fir Region." Pp. 3–7, in Chadwick Dearing Oliver, Donald P. Hanley, and Jay A. Johnson, eds. *Douglas-Fir: Stand Management for the Future.* Contribution no. 55. Institute of Forest Resources, University of Washington, Seattle, WA.

Lassoie, J. P., T. M. Hinckley, and C. C. Grier (1985). "Coniferous Forests of the Pacific Northwest." Pp. 127–61, in B. F. Chabot and H. A. Mooney, eds. *Physiological Ecology of North American Plant Communities.* Chapman and Hall, New York, NY.

Millar, Constance I. (1987). "The California Forest Germplasm Conservation Project: A Case for Genetic Conservation of Temperate Tree Species." *Conservation Biology* 1:191–93.

Mulcahy, David L. (1975). "Differential Mortality Among Cohorts in a Population of *Acer saccharum* (Aceraceae) Seedlings." *American Journal of Botany* 62:422–26.

Myren, Richard T., and Robert J. Ellis (1984). "Evapotranspiration in Forest Succession and Long-Term Effects on Fishery Resources: A Consideration for Management of Old-Growth Forests." Pp. 183–86, in William R. Meehan, Theodore R. Merrell, Jr., and Thomas A. Hanley, eds. *Fish and Wildlife Relationships in Old-Growth Forests*. American Institute of Fishery Research Biologists, Morehead City, NC.

Newman, E. I. (1985). "The Rhizosphere: Carbon Sources and Microbial Populations." Pp. 107–21, in A. H. Fitter, ed. *Ecological Interactions in Soil: Plants, Microbes and Animals*. Blackwell Scientific Publications, Oxford.

Olson, J. S., J. A. Watts, and L. J. Allison (1983). *Carbon in Live Vegetation of Major World Ecosystems*. TR004, U.S. Department of Energy, Washington, DC.

Pimentel, David, E. Garnick, A. Berkowitz, S. Jacobson, S. Napolitano, P. Black, S. Valdes-Cogliano, B. Vinzant, E. Hudes, and S. Littman (1980). "Environmental Quality and Natural Biota." *BioScience* 30(11):750–55.

Pritchett, William L., and Richard F. Fisher (1987). *Properties and Management of Forest Soils*. 2d ed. John Wiley & Sons, New York, NY.

Sedell, James R., and Frederick J. Swanson (1984). "Ecological Characteristics in Old-Growth Forests in the Pacific Northwest." Pp. 9–16, in William R. Meehan, Theodore R. Merrell, Jr., and Thomas A. Hanley, eds. *Fish and Wildlife Relationships in Old-Growth Forests*. American Institute of Fishery Research Biologists, Morehead City, NC.

Sheppard, Paul R., and Edward R. Cook (1988). "Scientific Value of Trees in Old-Growth Natural Areas." *Natural Areas Journal* 8(1):7–12.

Smith, William H. (1981). *Air Pollution and Forests: Interactions Between Air Contaminants and Forest Ecosystems*. Springer-Verlag, New York, NY.

Society of American Foresters (1984). *Scheduling the Harvest of Old Growth*. Society of American Foresters, Bethesda, MD.

Sollins, P., C. C. Grier, F. M. McCorison, K. Cromack, Jr., R. Fogel, and R. L. Fredriksen (1980). "The Internal Element Cycles of an Old-Growth Douglas-Fir Ecosystem in Western Oregon." *Ecological Monographs* 50(3):261–85.

Solomon, A. M., J. R. Trabalka, D. E. Reichle, and L. D. Voorhees (1985). "The Global Cycle of Carbon." Pp. 1–13, in J. R. Trabalka, ed. *Atmospheric Carbon Dioxide and the Global Carbon Cycle*. DOE/ER-0239, U.S. Department of Energy, Washington, DC.

Swanson, F. J., R. L. Fredriksen, and F. M. McCorison (1982). "Material Transfer in a Western Oregon Forested Watershed." Pp. 233–66, in Robert L. Edmonds, ed. *Analysis of Coniferous Forest Ecosystems in the Western United States*. Hutchinson Ross Publishing Co., Stroudsberg, PA.

Waring, R. H., and J. F. Franklin (1979). "Evergreen Coniferous Forests of the Pacific Northwest." *Science* 204:1380–86.

Waring, Richard H., and William H. Schlesinger (1985). *Forest Ecosystems, Concepts and Management*. Academic Press, Inc., Orlando, FL.

Westman, Walter E. (1977). "How Much Are Nature's Services Worth?" *Science* 197:960–64.

Wildlife Society, The (1988). *Management and Conservation of Old-Growth Forests in the U.S.—Position Statement*. The Wildlife Society, Washington, DC.

CHAPTER 6

Agee, James K. (1981). "Fire Effects on Pacific Northwest Forests: Flora, Fuels, and Fauna." Pp. 54–66, in *Northwest Fire Council 1981 Conference Proceedings.* Portland, OR.

Agee, James K., and Mark H. Huff (1987). "Fuel Succession in a Western Hemlock/Douglas-Fir Forest." *Canadian Journal of Forest Research* 17(7):697–704.

Amaranthus, Michael P., and David A. Perry (1989). "Interaction Effects of Vegetation Type and Pacific Madrone Soil Inocula on Survival, Growth, and Mycorrhiza Formation of Douglas-Fir." *Canadian Journal of Forest Research* 19(5):550–56.

Anderson, H. Michael, and Craig Gehrke (1988). *National Forests: Policies for the Future.* Vol. 1: *Water Quality and Timber Management.* The Wilderness Society, Washington, DC.

Binkley, Dan (1986). *Forest Nutrition Management.* John Wiley & Sons, New York, NY.

Brown, George W. (1980). *Forestry and Water Quality.* Oregon State University Book Stores, Inc., Corvallis, OR.

Cafferata, Stephen L. (1986). "Douglas-Fir Stand Establishment Overview: Western Oregon and Washington." Pp. 211–18, in Chadwick Dearing Oliver, Donald P. Hanley, and Jay A. Johnson, eds. *Douglas-Fir: Stand Management for the Future.* Contribution no. 55. Institute of Forest Resources, University of Washington, Seattle, WA.

California Gene Resource Program (1982). *Douglas-Fir Genetic Resources: An Assessment and Plan for California.* Prepared for the State of California under the administration of the Department of Food and Agriculture, Contract no. 9146. National Council on Gene Resources, Berkeley, CA.

Campbell, Robert K. (1979). "Genecology of Douglas-Fir in a Watershed in the Oregon Cascades." *Ecology* 60:1036–50.

Christiansen, Erik, Richard H. Waring, and Alan A. Berryman (1987). "Resistance of Conifers to Bark Beetle Attack: Searching for General Relationships." *Forest Ecology and Management* 22(1–2):89–106.

Conard, Susan G., Annabelle E. Jaramillo, Kermit Cromack, Jr., and Sharon Rose, eds. (1985). *The Role of the Genus* Ceanothus *in Western Forest Ecosystems.* USDA Forest Service, Pacific Northwest Forest and Range Experimental Station General Technical Report PNW-182, Portland, OR.

Corn, Paul Stephen, and R. Bruce Bury (1989). "Logging in Western Oregon: Responses of Headwater Habitats and Stream Amphibians." *Forest Ecology and Management* 29(01):1–19.

Coulson, Robert N., and John A. Witter (1984). *Forest Entomology: Ecology and Management.* John Wiley & Sons, New York, NY.

Diamond, Jared (1975). "The Island Dilemma: Lessons of Modern Biogeographic Studies for the Design of Nature Reserves." *Biological Conservation* 7:129–46.

Feller, M. C. (1982). *The Ecological Effects of Slashburning with Particular Reference to British Columbia: A Literature Review.* Province of British Columbia, Ministry of Forests, Land Management Report no. 13. Victoria, British Columbia.

Forman, R. T. T., and M. Godron (1986). *Landscape Ecology.* John Wiley & Sons, New York, NY.

Franklin, Jerry F., and Richard T. T. Forman (1987). "Creating Landscape Patterns by Forest Cutting: Ecological Consequences and Principles." *Landscape Ecology* 1(1):5–18.

Franklin, Jerry F., and Miles A. Hemstrom (1981). "Aspects of Succession in the Coniferous Forests of the Pacific Northwest." Pp. 212–29, in Darrell C. West, Herman H. Shugart, and Daniel B. Botkin, eds. *Forest Succession: Concepts and Application.* Springer-Verlag, New York, NY.

Franklin, Jerry F., H. H. Shugart, and Mark E. Harmon (1987). "Tree Death as an Ecological Process." *BioScience* 37(8):550–56.

Franklin, Jerry F., and Richard H. Waring (1980). "Distinctive Features of the Northwestern Coniferous Forest: Development, Structure, and Function." Pp. 59–85, in Richard H. Waring, ed. *Forests: Fresh Perspectives from Ecosystem Analysis.* Oregon State University Press, Corvallis, OR.

Fredriksen, Richard L., and R. Dennis Harr (1981). "Soil, Vegetation, and Watershed Management." Pp. 231–60, in Paul E. Heilman, Harry W. Anderson, and David M. Baumgartner, eds. *Forest Soils of the Douglas-Fir Region.* Washington State University, Pullman, WA.

Gara, Robert I. (1982). "Insect Pests of True Firs in the Pacific Northwest." Pp. 157–60, in Chadwick Dearing Oliver and Reid M. Kenady, eds. *Proceedings of the Biology and Management of True Fir in the Pacific Northwest Symposium.* Contribution no. 45. University of Washington College of Forest Resources, Seattle, WA.

Geppert, Rollin R., Charles W. Lorenz, and Arthur G. Larson (1984). *Cumulative Effects of Forest Practices on the Environment: A State of the Knowledge.* Ecosystems, Inc., Olympia, WA.

Graham, Joseph N., Edward W. Murray, and Don Minore (1982). *Environment, Vegetation, and Regeneration After Timber Harvest in the Hungry-Pickett Area of SW Oregon.* USDA Forest Service, Pacific Northwest Forest and Range Experiment Station Research Note PNW-400, Portland, OR.

Hadfield, James S. (1988). "Integrated Pest Management of a Western Spruce Budworm Outbreak in the Pacific Northwest." *Northwest Environmental Journal* 4(2):301–12.

Hansen, Everett M. (1977). "Forest Pathology: Fungi, Forests and Man." Pp. 107–24, in Tony Walters, ed. *Mushrooms and Man: An Interdisciplinary Approach to Mycology.* Lynn-Benton Community College, Albany, OR.

Hansen, E. M., D. J. Goheen, P. F. Hessburg, J. J. Witcosky, and T. D. Schowalter (1988). "Biology and Management of Black-Stain Root Disease in Douglas-Fir." Pp. 63–80, in T. C. Harrington and F. W. Cobb, eds. *Leptographium Root Diseases on Conifers.* American Phytopathological Society Press, St. Paul, MN.

Hardin, Garrett (1968). "The Tragedy of the Commons." *Science* 162:1243–48.

Harris, Larry D. (1984). *The Fragmented Forest: Island Biogeography Theory and the Preservation of Biotic Diversity.* University of Chicago Press, Chicago, IL.

Heath, B., P. Sollins, D. A. Perry, and K. Cromack, Jr. (1988). "Asymbiotic Nitrogen Fixation in Litter from Pacific Northwest Forests." *Canadian Journal of Forest Research* 18(1):68–74.

Hemstrom, Miles A., and Jerry F. Franklin (1982). "Fire and Other Disturbances of the Forests in Mount Rainier National Park." *Quaternary Research* 18(1):32–51.

Hewlett, John D. (1982). *Principles of Forest Hydrology.* University of Georgia Press, Athens, GA.

Heybroek, Hans M. (1982). "Monoculture Versus Mixture: Interactions Between Susceptible and Resistant Trees in a Mixed Stand." Pp. 326–41, in H. M. Heybroek, B. R. Stephan, and K. von Weissenberg, eds. *Resistance to Diseases and Pests in Forest Trees.* Centre for Agricultural Publishing and Documentation, Wageningen, the Netherlands.

Kimmins, J. P. (1987). *Forest Ecology.* Macmillan Publishing Co., New York, NY.

Likens, G. E., F. H. Bormann, R. S. Pierce, J. S. Eaton, and N. M. Johnson (1977). *Biogeochemistry of a Forest Ecosystem.* Springer-Verlag, New York, NY.

Little, Susan N., and Janet L. Ohmann (1988). "Estimating Nitrogen Lost from Forest Floor During Prescribed Fires in Douglas-Fir/Western Hemlock Clearcuts." *Forest Science* 31(1):152–64.

Lovejoy, T. E., R. O. Bierregaard, Jr., A. B. Rylands, J. R. Malcolm, C. E. Quintela, L. H. Harper, K. S. Brown, Jr., A. H. Powell, G. V. N. Powell, H. O. R. Schubart, and M. B. Hays (1986). "Edge and Other Effects of Isolation on Amazon Forest Fragments." Pp. 257–85, in Michael E. Soule, ed. *Conservation Biology: The Science of Scarcity and Diversity.* Sinauer Associates, Sunderland, MA.

MacArthur, R. H., and J. W. MacArthur (1961). "On Bird Species and Diversity." *Ecology* 42:594–98.

MacArthur, Robert H., and Edward O. Wilson (1967). *The Theory of Island Biogeography.* Princeton University Press, Princeton, NJ.

McDonald, G. I. (1981). *Differential Defoliation of Neighboring Douglas-Fir Trees by Western Spruce Budworm.* USDA Forest Service, Research Note INT-306. Ogden, UT.

Mann, L. K., D. W. Johnson, D. C. West, D. W. Cole, J. W. Hornbeck, C. W. Martin, H. Riekerk, C. T. Smith, W. T. Swank, L. M. Tritton, and D. H. Van Lear (1988). "Effects of Whole-Tree and Stem-Only Clearcutting on Postharvest Hydrologic Losses, Nutrient Capital, and Regrowth." *Forest Science* 34(2):412–28.

Means, Joseph E. (1982). "Developmental History of Dry Coniferous Forests in the Central Western Cascade Range of Oregon." Pp. 142–53, in Joseph E. Means, ed. *Forest Succession and Stand Development Research in the Northwest.* Forest Research Laboratory, Oregon State University, Corvallis, OR.

Morrison, Michael L., and E. Charles Meslow (1983). *Avifauna Associated with Vegetation on Clearcuts in the Oregon Coast Ranges.* USDA Forest Service, Pacific Northwest Forest and Range Experiment Station Research Paper PNW-305, Portland, OR.

Morrison, Peter H. (1984). "The History and Role of Fire in Forest Ecosystems of the Central Western Cascades of Oregon Determined by Forest Stand Analysis." M.S. thesis, University of Washington, Seattle, WA.

Newmark, William D. (1986). "Species-Area Relationship and Its Determinants for Mammals in Western North American National Parks." Pp. 83–98, in L. R. Heany and B. D. Patterson, eds. *Island Biogeography of Mammals.* Academic Press, Inc., Orlando, FL.

Norse, Elliott A., and Roger E. McManus (1980). "Ecology and Living Resources: Biological Diversity." Pp. 31–80, in *Environmental Quality 1980: The*

Eleventh Annual Report of the Council on Environmental Quality, Washington, DC.

Paine, R. T. (1966). "Food Web Complexity and Species Diversity." *American Naturalist* 100:65–75.

Perry, D. A. (1988). "Landscape Pattern and Forest Pests." *Northwest Environmental Journal* 4(2):213–28.

Petersen, T. D., M. Newton, and S. M. Zedaker (1988). "Influence of *Ceanothus velutinus* and Associated Forbs on the Water Stress and Stemwood Production of Douglas-Fir." *Forest Science* 34(2):333–43.

Price, Mary, and Nicholas Wasser (1979). "Pollen Dispersal and Optimal Outcrossing in *Delphinium nelsoni*." *Nature* 277:294–97.

Pyne, Stephen J. (1984). *Introduction to Wildland Fire: Fire Management in the United States*. John Wiley & Sons, New York, NY.

Raphael, Martin G. (1988). "Long-Term Trends in Abundance of Amphibians, Reptiles, and Mammals in Douglas-Fir Forests of Northwestern California." Pp. 23–31, in Robert C. Szaro, Kieth R. Severson, and David R. Patton, technical coordinators. *Management of Amphibians, Reptiles, and Small Mammals in North America*. USDA Forest Service, General Technical Report RM-166. Rocky Mountain Forest and Range Experiment Station, Ft. Collins, CO.

Raphael, Martin G., Kenneth V. Rosenberg, and Bruce G. Marcot (1988). "Large-Scale Changes in Bird Populations of Douglas-Fir Forests, Northwestern California." Pp. 63–83, in Jerome A. Jackson, ed. *Bird Conservation 3*. University of Wisconsin Press, Madison, WI.

Rehfeldt, G. E. (1983). "Genetic Variability Within Douglas-Fir Populations: Implications for Tree Improvement." *Silvae Genetica* 32:9–14.

Reid, Leslie M., and Thomas Dunne (1984). "Sediment Production from Forest Road Surfaces." *Water Resources Research* 20(11):1753–61.

Riggs, Lawrence A. (1988). *Reforestation Practices and Possible Genetic Consequences on National Forest Lands in the United States*. Report to The Wilderness Society, Washington, DC.

Robinson, Gordon (1988). *The Forest and the Trees: A Guide to Excellent Forestry*. Island Press, Washington, DC.

Root, R. B. (1973). "Organization of a Plant-Arthropod Association in Simple and Diverse Habitat: The Fauna of Collards." *Ecological Monographs* 43: 95–124.

Rosenberg, Kenneth V., and Martin G. Raphael (1986). "Effects of Forest Fragmentation on Vertebrates in Douglas-Fir Forests." Pp. 263–72, in Jared Verner, Michael L. Morrison, and John C. Ralph, eds. *Wildlife 2000: Modeling Habitat Relationships of Terrestrial Vertebrates*. University of Wisconsin Press, Madison, WI.

Schowalter, T. D. (1989). "Canopy Arthropod Community Structure and Herbivory in Old-Growth and Regenerating Forest in Western Oregon." *Canadian Journal of Forest Research* 19:318–22.

Schowalter, T. D., and J. E. Means (1988). "Pest Response to Simplification of Forest Landscapes." *Northwest Environmental Journal* 4(2):342–43.

Seastedt, T. R., and D. A. Crossley, Jr. (1981). "Microarthropod Response Following Cable Logging and Clear-Cutting in the Southern Appalachians." *Ecology* 62(1):126–35.

Sedell, James R., and Wayne S. Duval (1985). *Influence of Forest and Rangeland*

Management on Anadromous Fish Habitat in Western North America: Water Transportation and Storage of Logs. USDA Forest Service, Pacific Northwest Forest and Range Experiment Station General Technical Report PNW-186, Portland, OR.

Silvester, Warwick B., Phillip Sollins, Thomas Verhoven, and Steven P. Cline (1982). "Nitrogen Fixation and Acetylene Reduction in Decaying Conifer Boles: Effects of Incubation Time, Aeration, and Moisture Content." Canadian Journal of Forest Research 12(3):646–52.

Smith, David M. (1986). The Practice of Silviculture. John Wiley & Sons, New York, NY.

Sollins, P., and F. M. McCorison (1981). "Nitrogen and Carbon Solution Chemistry of an Old-Growth Coniferous Forest Watershed Before and After Cutting." Water Resources Research 17(5):1409–18.

Sollins, Phillip, Steven P. Cline, Thomas Verhoeven, Donald Sachs, and Gody Spycher (1987). "Patterns of Log Decay in Old-Growth Douglas-Fir Forests." Canadian Journal of Forest Research 17(12):1585–95.

Soule, Michael E., Douglas T. Bolger, Allison C. Alberts, John Wright, Marina Sorice, and Scott Hill (1988). "Reconstructed Dynamics of Rapid Extinctions of Chaparral-Requiring Birds in Urban Habitat Islands." Conservation Biology 2(1):75–92.

Stevenson, F. J. (1986). Cycles of Soil: Carbon, Nitrogen, Phosphorous, Sulfur, Micronutrients. John Wiley & Sons, New York, NY.

Stewart, R. E. (1978). "Site Preparation." Pp. 99–129, in Brian D. Cleary, Robert D. Greaves, and Richard K. Hermann, eds. Regenerating Oregon's Forests: A Guide for the Regeneration Forester. Oregon State University Extension Service, Corvallis, OR.

Swanston, Doug, ed. (1985). Proceedings of a Workshop on Slope Stability: Problems and Solutions in Forest Management. USDA Forest Service, Pacific Northwest Forest and Range Experiment Station General Technical Report PNW-180. Portland, OR.

Terborgh, John (1974). "Preservation of Natural Diversity: The Problem of Extinction Prone Species." BioScience 24:715–22.

Walstad, John D., Michael Newton, and Raymond J. Boyd, Jr. (1987). "Forest Vegetation Problems in the Northwest." Pp. 15–53, in John D. Walstad and Peter J. Kuch, eds. Forest Vegetation Management for Conifer Production. John Wiley & Sons, New York, NY.

Williamson, Richard L., and Asa D. Twombly (1983). "Pacific Douglas-Fir." Pp. 9–12, in Russel M. Burns, ed. Silvicultural Systems for the Major Forest Types of the United States. USDA Forest Service Agriculture Handbook no. 445. Washington, DC.

Wright, H. E., Jr., and M. L. Heinselman (1973). "Introduction—the Ecological Role of Fire in Natural Conifer Forests of Western and Northern North America." Quaternary Research 3(3):319–28.

Youngberg, Chester T. (1978). Forest Soils and Land Use: Proceedings of the Fifth North American Forest Soils Conference, Colorado State University, Ft. Collins, Colorado, August, 1978. Colorado State University, Ft. Collins, CO.

Zobel, Donald B., Lewis F. Roth, and Glenn M. Hawk (1985). Ecology, Pathology and Management of Port-Orford-Cedar (Chamaecyparis lawsoniana). USDA Forest Service, Pacific Northwest Forest and Range Experiment Station General Technical Report PNW-184, Portland, OR.

CHAPTER 7

Abrahamson, Dean Edwin, ed. (1989). *The Challenge of Global Warming*. Island Press, Washington, DC.

Bolin, Bert, et al., eds. (1986). *The Greenhouse Effect, Climatic Change, and Ecosystems*. John Wiley & Sons, New York, NY.

Clark, James S. (1988). "Effect of Climatic Change on Fire Regimes in Northwestern Minnesota." *Nature* 334:233–35.

Cwynar, Les C. (1987). "Fire and the Forest History in the North Cascade Range." *Ecology* 68(4):791–802.

Davis, Margaret Bryan (1981). "Quaternary History and the Stability of Forest Communities." Pp. 132–53, in Darrell C. West, Herman H. Shugart, and Daniel B. Botkin, eds. *Forest Succession: Concepts and Application*. Springer-Verlag, New York, NY.

Denison, William C., and Sue M. Carpenter (1973). *A Guide to Air Quality Monitoring with Lichens*. Lichen Technology, Inc., Corvallis, OR.

Dickinson, Robert E., and Ralph J. Cicerone (1986). "Future Global Warming from Atmospheric Trace Gases." *Nature* 319:109–15.

Edmonds, R. L., and F. A. Basabe (1987). "Ozone and Acidic Inputs into Forests in Northwest USA." *EXS 51: Advances in Aerobiology*, pp. 319–24.

Emanuel, William R., Herman H. Shugart, and Mary P. Stevenson (1985). "Climatic Change and the Broad-Scale Distribution of Terrestrial Ecosystem Complexes." *Climatic Change* 7(1):29–43.

Franklin, Jerry F., and C. T. Dyrness (1973). *Natural Vegetation of Oregon and Washington*. USDA Forest Service, Pacific Northwest Forest and Range Experiment Station General Technical Report PNW-8. Portland, OR.

Franklin, Jerry F., and Richard T. T. Forman (1987). "Creating Landscape Patterns by Forest Cutting: Ecological Consequences and Principles." *Landscape Ecology* 1(1):5–18.

Guderian, Robert, ed. (1985). *Air Pollution by Photochemical Oxidants: Formation, Transport, Control, and Effects on Plants*. Springer-Verlag, Berlin.

Hadfield, James S. (1988). "Integrated Pest Management of a Western Spruce Budworm Outbreak in the Pacific Northwest." *Northwest Environmental Journal* 4(2):301–12.

Henderson, Jan, and Linda Brubaker (1986). "Responses of Douglas-Fir to Long-Term Variations in Precipitation and Temperature in Western Washington." Pp. 162–67, in Chadwick Dearing Oliver, Donald P. Hanley, and Jay A. Johnson, eds. *Douglas-Fir: Stand Management for the Future*. Contribution no. 55. Institute of Forest Resources, University of Washington, Seattle, WA.

Janzen, Daniel H. (1986). "The Eternal External Threat." Pp. 286–303, in Michael E. Soule, ed. *Conservation Biology: The Science of Scarcity and Diversity*. Sinauer Associates, Sunderland, MA.

Kossuth, Susan V., and R. Hilton Biggs (1981). "Ultraviolet-B Radiation Effects on Early Seedling Growth of Pinaceae Species." *Canadian Journal of Forest Research* 11:243–48.

Kramer, Paul J., and Nasser Sionit (1987). "Effects of Increasing Carbon Dioxide Concentration on the Physiology and Growth of Forest Trees." Pp. 219–46, in William S. Shands and John S. Hoffman, eds. *The Greenhouse Effect,*

Climate Change, and U.S. Forests. The Conservation Foundation, Washington, DC.

Kruckeberg, Arthur R. (1984). California Serpentines: Flora, Vegetation, Geology, Soils and Management Problems. University of California Press, Berkeley, CA.

Kulp, J. Laurence (1986). "Projections for the Year 2020 in the Douglas-Fir Region." Pp. 3–7, in Chadwick Dearing Oliver, Donald P. Hanley, and Jay A. Johnson, eds. Douglas-Fir: Stand Management for the Future. Contribution no. 55. Institute of Forest Resources, University of Washington, Seattle, WA.

Lemon, E. R., ed. (1983). CO_2 and Plants—the Response of Plants to Rising Levels of Atmospheric Carbon Dioxide. Westview Press, Boulder, CO.

Leverenz, Jerry W., and Deborah J. Lev (1987). "Effects of Carbon Dioxide-Induced Climate Changes on the Natural Ranges of Six Major Commercial Tree Species in the Western United States." Pp. 123–55, in William S. Shands and John S. Hoffman, eds. The Greenhouse Effect, Climate Change, and U.S. Forests. The Conservation Foundation, Washington, DC.

Lincoln, D. E., N. Sionit, and B. R. Strain (1984). "Growth and Feeding Response of Psuedoplusia includens (Lepidoptera: Noctuidae) to Host Plants Grown in Controlled Carbon Dioxide Atmospheres." Environmental Entomology 13(6):1527–30.

Nyberg, J. Brian, Alton S. Harestad, and Fred L. Bunnell (1987). " 'Old Growth' by Design: Managing Young Forests for Old-Growth Wildlife." Pp. 70–81, in Richard E. McCabe, ed. Transactions of the Fifty-Second North American Wildlife and Natural Resources Conference. Wildlife Management Institute, Washington, DC.

Perry, D. A., M. P. Amaranthus, J. G. Borchers, S. L. Borchers, and R. E. Brainerd (1989). "Bootstrapping in Ecosystems." BioScience 39(4):230–37.

Perry, David A., and Jumanne Maghembe (1989). "Ecosystem Concepts and Current Trends in Forest Management: Time for Reappraisal." Forest Ecology and Management 26:123–40.

Perry, D. A., and G. B. Pitman (1983). "Genetic and Environmental Influences in Host Resistance to Herbivory: Douglas-Fir and the Western Spruce Budworm." Zeitschrift Fuer Angewandte Entomologie 96:217–28.

Peters, Robert L., and Joan S. Darling (1985). "The Greenhouse Effect and Nature Reserves." BioScience 35(11):707–17.

Schowalter, T. D., and J. E. Means (1988). "Pest Response to Simplification of Forest Landscapes." Northwest Environmental Journal 4(2):342–43.

Strain, Boyd R. (1987). "Direct Effects of Increasing Atmospheric CO_2 on Plants and Ecosystems." Trends in Ecology and Evolution 2(1):18–21.

Sullivan, Joe H., and Alan H. Teramura (1988). "Effects of Ultraviolet-B Irradiation on Seedling Growth in the Pinaceae." American Journal of Botany 75(2):225–30.

USDA Forest Service (1988). A Guide to Sensitive Plants of the Siskiyou National Forest. USDA Forest Service, Siskiyou National Forest, Grants Pass, OR.

Waring, R. H., and J. F. Franklin (1979). "Evergreen Coniferous Forests of the Pacific Northwest." Science 204:1380–86.

Woodman, James N. (1987). "Potential Impact of Carbon Dioxide-Induced Climate Changes on Management of Douglas-Fir and Western Hemlock." Pp. 277–83, in William S. Shands and John S. Hoffman, eds. The Greenhouse

Effect, Climate Change, and U.S. Forests. The Conservation Foundation, Washington, DC.

World Climate Programme (1988). *Developing Policies for Responding to Climatic Change.* WCIP-1, WMO/TD no. 225, Beijer Institute, Stockholm.

CHAPTER 8

Brown, E. Reade, and Alan B. Curtis (1985). "Introduction." Pp. 1–15, in E. Reade Brown, ed. *Management of Wildlife and Fish Habitats in Forests of Western Oregon and Washington, Part 1—Chapter Narratives,* USDA Forest Service, Pacific Northwest Region, Portland, OR.

DeBell, Dean S., and Jerry F. Franklin (1987). "Old-Growth Douglas-Fir and Western Hemlock: A 36-Year Record of Growth and Mortality." *Western Journal of Applied Forestry* 2(4):111–14.

Findley, Rowe (1981). "Eruption of Mount St. Helens." *National Geographic* 159(1):2–65.

Franklin, Jerry F., and C. T. Dyrness (1973). *Natural Vegetation of Oregon and Washington.* USDA Forest Service, Pacific Northwest Forest and Range Experiment Station General Technical Report PNW-8, Portland, OR.

Franklin, Jerry F., H. H. Shugart, and Mark E. Harmon (1987). "Tree Death as an Ecological Process." *BioScience* 37(8):550–56.

Greene, Sarah (1988). "Research Natural Areas and Protecting Old-Growth Forests on Federal Lands in Western Oregon and Washington." *Natural Areas Journal* 8(1):25–30.

Harris, Larry D. (1984). *The Fragmented Forest: Island Biogeography Theory and the Preservation of Biotic Diversity.* University of Chicago Press, Chicago, IL.

Haynes, Richard W. (1986). *Inventory and Value of Old-Growth in the Douglas-Fir Region.* USDA Forest Service, Pacific Northwest Research Station Research Note PNW-437, Portland, OR.

Leopold, Aldo (1949). "The Round River." P. 90, in *A Sand County Almanac, with Essays on Conservation from Round River.* Oxford University Press, New York, NY.

Maser, Chris (1988). *The Redesigned Forest.* R&E Miles, San Pedro, CA.

Morrison, Peter H. (1988). *Old Growth in the Pacific Northwest: A Status Report.* The Wilderness Society, Washington, DC.

Old-Growth Definition Task Group (J. F. Franklin, F. Hall, W. Laudenslayer, C. Maser, J. Nunan, J. Poppino, C. J. Ralph, and T. Spies) (1986). *Interim Definitions for Old-Growth Douglas-Fir and Mixed-Conifer Forests in the Pacific Northwest and California.* USDA Forest Service, Pacific Northwest Research Station Research Note PNW-447, Portland, OR.

O'Toole, Randal (1988). *Reforming the Forest Service.* Island Press, Washington, DC.

Robinson, Gordon (1988). *The Forest and the Trees: A Guide to Excellent Forestry.* Island Press, Washington, DC.

Society of American Foresters (1984). *Scheduling the Harvest of Old Growth.* Society of American Foresters, Bethesda, MD.

Spies, Thomas A., and Jerry F. Franklin (1988). "Old Growth and Forest Dynamics in the Douglas-Fir Region of Western Oregon and Washington." *Natural Areas Journal* 8(3):190–201.

Thomas, Jack Ward, Leonard F. Ruggiero, R. William Mannan, John W. Schoen, and Richard A. Lancia (1988). "Management and Conservation of Old-Growth Forests in the United States." *Wildlife Society Bulletin* 16:252–62.

CHAPTER 9

Norse, Elliott A., Kenneth L. Rosenbaum, David S. Wilcove, Bruce A. Wilcox, William H. Romme, David W. Johnston, and Martha L. Stout (1986). *Conserving Biological Diversity in Our National Forests*. The Wilderness Society, Washington, DC.

Old-Growth Definition Task Group (J. F. Franklin, F. Hall, W. Laudenslayer, C. Maser, J. Nunan, J. Poppino, C. J. Ralph, and T. Spies) (1986). *Interim Definitions for Old-Growth Douglas-Fir and Mixed-Conifer Forests in the Pacific Northwest and California*. USDA Forest Service, Pacific Northwest Research Station Research Note PNW-447, Portland, OR.

Vitousek, Peter M., Paul R. Ehrlich, Anne H. Ehrlich, and Pamela A. Matson (1986). "Human Appropriation of the Products of Photosynthesis." *BioScience* 36(6):368–73.

Whittaker, R. H. (1960). "Vegetation of the Siskiyou Mountains, Oregon and California." *Ecological Monographs* 30:279–338.

Wilcove, David S. (1988). *National Forests: Policies for the Future*. Vol. 2: *Protecting Biological Diversity*. The Wilderness Society, Washington, DC.

Wilderness Society, The (1988). *New Directions for the Forest Service*. The Wilderness Society, Washington, DC.

Index

About the Author

Since 1987 Elliott A. Norse has been senior ecologist for The Wilderness Society. Previously, he was public policy director for The Ecological Society of America and staff ecologist for the President's Council on Environmental Quality. A coauthor of *Conserving Biological Diversity in Our National Forests*, published by The Wilderness Society in 1986, Dr. Norse has also contributed to a wide range of scientific journals. In 1980, for the Council on Environmental Quality, he was coauthor of a paper that was instrumental in defining the term *biological diversity* and identifying the alarming rate of extinctions. Dr. Norse lives in Washington, D.C.

Also Available from Island Press

The Challenge of Global Warming
Edited by Dean Edwin Abrahamson

The Forest and the Trees: A Guide to Excellent Forestry
By Gordon Robinson

Land and Resource Planning in the National Forests
By Charles F. Wilkinson and H. Michael Anderson

Last Stand of the Red Spruce
By Robert A. Mello

Our Common Lands: Defending the National Parks
Edited by David J. Simon

Pocket Flora of the Redwood Forest
By Dr. Rudolf W. Becking

Reforming the Forest Service
By Randal O'Toole

Saving the Tropical Forests
By Judith Gradwohl and Russell Greenberg

The Sierra Nevada: A Mountain Journey
By Tim Palmer

Shading Our Cities: A Resource Guide for Urban and Community Forests
Edited by Gary Moll and Sara Ebenreck

Tree Talk: The People and Politics of Timber
By Ray Raphael

For a complete catalog of Island Press publications, please write:
Island Press
Box 7
Covelo, CA 95428

Island Press Board of Directors